HBase 入门与实践

彭旭 著

人民邮电出版社

北京

图书在版编目（CIP）数据

HBase入门与实践 / 彭旭著． -- 北京：人民邮电出版社，2018.12（2021.8重印）
ISBN 978-7-115-49383-5

Ⅰ．①H… Ⅱ．①彭… Ⅲ．①计算机网络-信息存贮 Ⅳ．①TP393

中国版本图书馆CIP数据核字(2018)第214274号

内 容 提 要

本书以精练的语言介绍 HBase 的基础知识，让初学者能够快速上手使用 HBase，对 HBase 的核心思想（如数据读取、数据备份等）和 HBase 架构（如 LSM 树、WAL）有深入的分析，并且让有经验的 HBase 开发人员也能够循序渐进地深入理解 HBase 源码，以便更好地去调试和解决线上遇到的各种问题。本书更加专注于 HBase 在线实时系统的调优，使 HBase 集群响应延迟更低。本书结合企业必备的"用户行为分析系统"，让读者能够快速上手的同时，也不乏企业 HBase 实际应用场景，理论不脱离实际，真正做到从入门到精通。

本书适合有一定 Java 基础的程序员作为 HBase 入门教程，HBase 运维人员可以将本书作为参考手册来部署和监控 HBase，正在将 HBase 应用到在线生产环境中的软件开发人员也可以参考本书来调优 HBase 在线集群性能。

◆ 著　　彭　旭
　责任编辑　杨海玲
　责任印制　焦志炜

◆ 人民邮电出版社出版发行　北京市丰台区成寿寺路 11 号
　邮编　100164　电子邮件　315@ptpress.com.cn
　网址　http://www.ptpress.com.cn
　北京虎彩文化传播有限公司印刷

◆ 开本：800×1000　1/16
　印张：14
　字数：292 千字　　　　　　2018 年 12 月第 1 版
　印数：6 801 – 7 300 册　　2021 年 8 月北京第 10 次印刷

定价：59.00 元

读者服务热线：(010)81055410　印装质量热线：(010)81055316
反盗版热线：(010)81055315
广告经营许可证：京东市监广登字 20170147 号

对本书的赞誉

本书循序渐进地介绍了 HBase 从入门到企业实践调优,深入浅出地阐述了 HBase 架构与实现原理,以企业必备的用户行为系统为实践场景,用通俗易懂的语言带领读者走进 HBase 世界。作者在魅族云和大数据充分实践了书中架构,并实际带来了质量和效率的提升,同时也降低了云端运营成本。

——李柯辰,前魅族平台事业部总经理,现卓轩科技 CEO

彭旭是一个优秀的程序员、架构师。本书在介绍 HBase 基础知识的同时融入了他在魅族云服务团队将存储系统迁移到 HBase 的经验和教训,此外还在 HBase 源码和架构研究上有涉猎,对初学者、Java 相关开发人员和 HBase 运维人员等都是一本不错的参考书。

——何伟,前魅族 Flyme 技术委员会主席,现卓轩科技 CTO

作者将多年 HBase 的实践经验与心得体会沉淀为本书,既有经典的案例分析,也有抽丝剥茧的源码分析,对于大数据行业从业者,非常值得一读。

——张发恩,前百度云技术委员会主席,创新工场人工智能工程院首席架构师,创新奇智 CTO

作者是一个长期战斗在一线的典型程序员,本书所写的内容均贴近现实应用场景,用线上真实的案例来描述如何构建、优化 HBase 生成环境,事半功倍,不管是初学者还是入门者都可以从这些案例中吸取经验。

——唐进,前百度云产品委员会主席,现爱乐奇 CTO

与拖沓烦冗的英文著作相比,本书篇幅适中,但是以精简干练的语言完整地描述了 Hbase 从初学到熟练应用所需学习的所有 HBase 知识点。

——马杜,华云数据执行总裁

HBase 是一个历久弥坚的分布式列式存储系统,相关图书出版时间均距今已久。本书以当前稳定版本 HBase 为基础,着重介绍了 HBase 在线实时系统的调优等,很适合正在将 HBase 应用到在线生产环境的开发人员和运维人员阅读。2010 年下半年当当网在纽交所上市之前,我受当时 CTO 的委托加强当时的防盗刷/刷单系统,本质上是要升级其中的高速计数器产品模块,该模块持久化部分的技术选型就是 HBase。这是当当网技术第一次在生产环境中应用 Hadoop 和 HBase。在这个过程中,最让人记忆深刻的不是技术难度,而是当时的资料很有限,HBase 的技术迭代又很快,网上大量文档在不同程度上有过时的问题,与其在网站上找过时的资料,不如直接读代码。因此，本书全面介绍了当前稳定版本的HBase，真是为HBase相关开发人员提供了很大的便利。

——傅强，前当当网技术副总裁，现九枝兰合伙人

HBase是Apache下的项目，它是一个高可用、高性能、可伸缩和面向列的分布式存储系统，能实现对海量数据的高性能的读写。本书作者是前阿里天猫的工程师，有着丰富的开发经验，相信这本书一定能让读者的HBase理论与实践水平更上一层楼。

——杨开振，《深入浅出Spring Boot 2.x》《深入浅出MyBatis技术原理与实战》
《Java EE互联网轻量级框架整合开发：SSM框架（Spring MVC+Spring+MyBatis）
和Redis实现》作者

一个有很大价值的开源项目既需要活跃的代码贡献者为其快速迭代铺路，也需要有优质的入门资料为广大初学者打下良好的基础。要学习HBase，英文版的*Apache HBase Book*仍然是最佳的参考手册，而对英文不是特别好的读者来说，可以选择的资料就是《HBase权威指南》或者是一些技术作者的专栏博客。本书对入门者来说是很好的中文读物。书中配有丰富的案例和插图，让读者可以较为轻松地理解HBase的常用场景和用法。希望本书对广大的HBase爱好者有所裨益。

——胡争，小米HBase工程师，HBase Committer

我几年前在维护淘宝的HBase集群时，时常和业务方交流如何基于HBase的架构来设计一个可靠和高性能的业务系统，深感对HBase内部原理的了解能极大改善业务系统的性能、稳定性以及成本。本书以当前HBase最新的稳定版本为基础，十分精准地抽离出了开发者需要重点关注的特点，并辅以几个典型的场景加以分析，堪称业务系统设计的手边助理。此外，作者拥有丰富的一线经验，因此本书具有很高的帮助开发者迅速上手生产系统的安装、部署以及API使用的实践价值。

——邓明鉴，前阿里巴巴高级专家，现ZStack首席架构师

前　言

人工智能作为当前最热门的技术，其根本上离不开大数据的支持。如果把人工智能比喻成一个神经网络，那么数据则是在这个神经网络中用来传递信息的化学物质，没有信息传递的神经网络显然不名一文，因此大数据扮演着人工智能基石的角色。Hadoop 生态系统的 HDFS 和 MapReduce 分别为大数据提供了存储和分析处理能力，但是对在线实时的数据存取则无能为力，而 HBase 作为 Apache 顶级项目，弥补了 Hadoop 的这一缺陷，满足了在线实时系统低延时的需求。目前 HBase 在各大互联网公司几乎都有应用，前景广阔。

本书先介绍 HBase 基础知识，再带领读者深入研读 HBase 源码，从数据读取之 Scan 流程、HBase 架构（如 LSM 树和 WAL），到构建线上实时低延迟系统的调优，结合企业必备的"用户行为分析系统"，让读者能够快速上手 HBase 并了解 HBase 实现原理，同时让读者通过分析源码来了解 HBase 的设计思想，期望读者能够真正做到从入门到精通。

本书各章的主要内容概括如下。

- 第 1 章介绍 HBase 的诞生背景与 HBase 在国内各大公司的一些应用场景。
- 第 2 章介绍 HBase 的安装模式以及分布式部署情况下依赖 Hadoop 和 ZooKeeper 的安装。
- 第 3 章介绍与 HBase 逻辑模型和物理模型相关的概念，如表、分区、行键、列族等，以及这些模型物理上的存储结构。
- 第 4 章介绍 HBase shell，在 shell 中使用数据定义语言与数据操纵语言，包括如何创建表、修改表、操作表的数据、使用 Scan 过滤所需要查询的数据，还介绍了一些特殊的命令，如查看集群间复制状态、管理分区等。
- 第 5 章介绍如何将 HBase 作为存储系统，包括行键设计原则、选用宽表或者高表，最后以朋友圈为例来进一步介绍如何在应用中合理地设计 HBase 表。
- 第 6 章介绍 HBase Java 客户端 API，包括在 Java API 中使用数据定义语言与数据操作语言、过滤器的使用、HBase 事务支持。
- 第 7 章介绍 HBase 架构设计，包括从 B 树开始到如何使用 LSM 来提升写效率、WAL 的作用、数据写入流程、数据读取流程。
- 第 8 章介绍两种类型的协处理器，即观察者类型协处理器和端点类型协处理器，以及如何装载使用。
- 第 9 章介绍 HBase 在线调优，包括客户端层面如何设置查询缓存、跳过 WAL 写入，服务端层面建表语句优化、开启机柜感知、本地读、补偿重试读以及分区服务器 JVM 内存调优等。
- 第 10 章介绍数据备份与复制的几种方式，包括复制、快照、导入、导出和复制表。
- 第 11 章介绍 HBase 自带的监控系统，着重介绍了 HBase JMX 监控。

作为一个典型的内敛型程序员，我从来没有过要写书、当作者的念头，也一直觉得作家是一个很神秘的职业。由于个人性格问题，我对任何事情的描述都是直来直往，描述问题经常是直奔主题，很少会去解释背景与前因后果。机缘巧合下我遇到了几个做教育以及写书的朋友，一番纠结后兴起了写书的念头，正好最近一直在研究 HBase，加上已经到了三十而立的年龄，宝宝也将要出生，觉得自己也需要做点什么，一方面给自己留点回忆，另一方面等宝宝长大后也算是给她的一个礼物。开始写的时候进度很慢，感觉无法坚持下去，但是慢慢地写着写着竟然越来越精神，越来越投入，愿本书能够帮到你！

感谢我的父母对我个人和家庭无私的付出与无微不至的照顾，感谢我的妻子黄晶对我的包容与支持！感谢珠海市魅族科技有限公司、珠海市卓轩科技有限公司提供的平台与实践经历促成了本书！感谢李柯辰先生与何伟先生对我和本书的鼎力支持与帮助！感谢我生命中遇到的每一个人，愿大家越来越好！祝我的宝贝彭语桐开心快乐成长！

由于作者水平有限，在编写过程中，难免出现错误或者不准确的地方，但是凡是涉及实例代码之处，我都实际运行过程序，从而保证程序的准确性。如果读者在阅读过程中发现有错误之处，敬请指正，联系方式如下：

微博 17051158029
微信 17051158029
邮箱 17051158029@163.com

彭旭
2018 年 10 月

资源与支持

本书由异步社区出品，社区（https://www.epubit.com/）为您提供相关资源和后续服务。

配套资源

本书提供如下资源：
- 本书源代码；
- 书中彩图文件。

要获得以上配套资源，请在异步社区本书页面中单击 配套资源 ，跳转到下载界面，按提示进行操作即可。注意：为保证购书读者的权益，该操作会给出相关提示，要求输入提取码进行验证。

如果您是教师，希望获得教学配套资源，请在社区本书页面中直接联系本书的责任编辑。

提交勘误

作者和编辑尽最大努力来确保书中内容的准确性，但难免会存在疏漏。欢迎您将发现的问题反馈给我们，帮助我们提升图书的质量。

当您发现错误时，请登录异步社区，按书名搜索，进入本书页面，单击"提交勘误"，输入勘误信息，单击"提交"按钮即可。本书的作者和编辑会对您提交的勘误进行审核，确认并接受后，您将获赠异步社区的 100 积分。积分可用于在异步社区兑换优惠券、样书或奖品。

扫码关注本书

扫描下方二维码，您将会在异步社区微信服务号中看到本书信息及相关的服务提示。

与我们联系

我们的联系邮箱是 contact@epubit.com.cn。

如果您对本书有任何疑问或建议，请您发邮件给我们，并请在邮件标题中注明本书书名，以便我们更高效地做出反馈。

如果您有兴趣出版图书、录制教学视频，或者参与图书翻译、技术审校等工作，可以发邮件给我们；有意出版图书的作者也可以到异步社区在线提交投稿（直接访问www.epubit.com/selfpublish/submission 即可）。

如果您是学校、培训机构或企业，想批量购买本书或异步社区出版的其他图书，也可以发邮件给我们。

如果您在网上发现有针对异步社区出品图书的各种形式的盗版行为，包括对图书全部或部分内容的非授权传播，请您将怀疑有侵权行为的链接发邮件给我们。您的这一举动是对作者权益的保护，也是我们持续为您提供有价值的内容的动力之源。

关于异步社区和异步图书

"异步社区"是人民邮电出版社旗下 IT 专业图书社区，致力于出版精品 IT 技术图书和相关学习产品，为作译者提供优质出版服务。异步社区创办于 2015 年 8 月，提供大量精品 IT 技术图书和电子书，以及高品质技术文章和视频课程。更多详情请访问异步社区官网 https://www.epubit.com。

"异步图书"是由异步社区编辑团队策划出版的精品 IT 专业图书的品牌，依托于人民邮电出版社近 30 年的计算机图书出版积累和专业编辑团队，相关图书在封面上印有异步图书的 LOGO。异步图书的出版领域包括软件开发、大数据、AI、测试、前端、网络技术等。

异步社区

微信服务号

目 录

第 1 章 HBase 简介 1
1.1 背景 1
1.2 NoSQL 与传统 RDBMS 2
1.3 应用场景 3
1.3.1 Facebook 用户交互数据 3
1.3.2 淘宝 TLog 等 3
1.3.3 小米云服务 4
1.3.4 用户行为数据存储 4

第 2 章 HBase 安装 5
2.1 单机部署 5
2.1.1 前置条件 5
2.1.2 下载 HBase 6
2.1.3 配置 HBase 6
2.1.4 启动 HBase 6
2.1.5 HBase 初体验 7
2.2 分布式部署 8
2.2.1 环境准备 9
2.2.2 ZooKeeper 安装 11
2.2.3 Hadoop 安装 12
2.2.4 HBase 安装 23
2.2.5 启动集群 28
2.3 集群增删节点 29
2.3.1 增加节点 29
2.3.2 删除节点 30

第 3 章 HBase 数据模型 33
3.1 逻辑模型 33
3.2 物理模型 35

第 4 章 HBase shell 39
4.1 数据定义语言 39
4.1.1 创建表 39
4.1.2 查看所有表 40
4.1.3 查看建表 40
4.1.4 修改表 41
4.2 数据操纵语言 41
4.2.1 Put 41
4.2.2 Get 42
4.2.3 Scan 43
4.2.4 删除数据 45
4.3 其他常用 shell 46
4.3.1 复制状态查看 46
4.3.2 分区拆分 47
4.3.3 分区主压缩 47
4.3.4 负载均衡开关 48
4.3.5 分区手动迁移 48

第 5 章 模式设计 49
5.1 行键设计 50
5.2 规避热点区间 52
5.3 高表与宽表 54
5.4 微信朋友圈设计 55
5.4.1 需求定义 55
5.4.2 问题建模 55

第 6 章 客户端 API 61
6.1 Java 客户端使用 61
6.2 数据定义语言 64

	6.2.1	表管理 ·· 64
	6.2.2	分区管理 ·· 66
6.3	数据操纵语言 ·· 68
	6.3.1	Put ·· 68
	6.3.2	Get ·· 70
	6.3.3	Scan ·· 72
	6.3.4	Delete ·· 74
	6.3.5	Increment ·· 76
6.4	过滤器 ·· 78
	6.4.1	过滤器简介 ·· 78
	6.4.2	过滤器使用 ·· 81
6.5	事务 ·· 94
	6.5.1	原子性 ·· 95
	6.5.2	隔离性 ·· 95

第 7 章 架构实现 ·· 101

7.1	存储 ·· 101
	7.1.1 B+树 ·· 101
	7.1.2 LSM 树 ·· 102
	7.1.3 WAL ·· 104
7.2	数据写入读取 ·· 107
	7.2.1 定位分区服务器 ·· 107
	7.2.2 数据修改流程 ·· 108
	7.2.3 数据查询流程 ·· 113

第 8 章 协处理器 ·· 115

8.1	观察者类型协处理器 ·· 115
8.2	端点类型协处理器 ·· 132
8.3	装载/卸载协处理器 ·· 136
	8.3.1 静态装载/卸载 ·· 136
	8.3.2 动态装载/卸载 ·· 137

第 9 章 HBase 性能调优 ·· 141

9.1	客户端调优 ·· 141
	9.1.1 设置客户端写入缓存 ·· 141
	9.1.2 设置合适的扫描缓存 ·· 143

	9.1.3	跳过 WAL 写入 ·· 143
	9.1.4	设置重试次数与间隔 ·· 144
	9.1.5	选用合适的过滤器 ·· 144
9.2	服务端调优 ·· 145
	9.2.1	建表 DDL 优化 ·· 145
	9.2.2	禁止分区自动拆分与压缩 ·· 150
	9.2.3	开启机柜感知 ·· 151
	9.2.4	开启 Short Circuit Local Reads ·· 153
	9.2.5	开启补偿重试读 ·· 154
	9.2.6	JVM 内存调优 ·· 155

第 10 章 集群间数据复制 ·· 163

10.1	复制 ·· 164
	10.1.1 集群拓扑 ·· 165
	10.1.2 配置集群复制 ·· 166
	10.1.3 验证复制数据 ·· 169
	10.1.4 复制详解 ·· 171
10.2	快照 ·· 175
	10.2.1 配置快照 ·· 176
	10.2.2 管理快照 ·· 176
10.3	导出和导入 ·· 178
	10.3.1 导出 ·· 178
	10.3.2 导入 ·· 180
10.4	复制表 ·· 180

第 11 章 监控 ·· 183

11.1	Hadoop 监控 ·· 183
	11.1.1 Web 监控页面 ·· 184
	11.1.2 JMX 监控 ·· 185
11.2	HBase 监控 ·· 188
	11.2.1 Web 监控页面 ·· 188
	11.2.2 JMX 监控 ·· 190
	11.2.3 API 监控 ·· 192

后记 ·· 195

附录　常见问题 ·· 197

第 1 章

HBase 简介

HBase 至今已有十多年的历史,其分布式弹性伸缩的特性使其在今天的大数据存储系统中占有重要的一席之地。

1.1 背景

随着计算机科技技术的发展,电子设备越来越普及,价格也越来越便宜,各种电子设备(如手机、智能家居、网络摄像头、智能汽车等)产生的数据量正在急剧增长,全球数据量的迅猛增长为大数据行业的发展提供了基础保证。市场调研机构预计未来全球每年数据量的增长率将达到 40%以上,而且数据增长速度越来越快,对一些大公司来说,每天甚至每小时都可以产生 TB、PB 级的数据,到 2020 年,全球数据总量预计将超过 50 ZB。

这些数据可能很难存储或者分析,但是很多时候这些数据很有可能成为企业的第一生产力。例如,阿里巴巴的用户行为数据,对于推荐或者在线广告至关重要,如果你在淘宝搜索了一款手机,稍后当你浏览新闻的时候,新闻客户端就会马上给你推荐一些手机的广告,这样有针对性的广告也就是常说的千人千面,极大地提高了广告的转化率。

在分布式数据库系统发展之前,因为缺乏有效或者低成本的方式来存储大量的数据,很多企业会忽略某些数据源,这样导致当真正需要使用某些数据的时候已经找不到可以用来筛选的源数据了,或者即使使用低成本的方式(如廉价磁带或者磁盘)将数据存储下来了,也因为缺乏有效的数据筛选等处理手段而导致数据无法发挥其应有的价值。

传统系统可能很难存储、分析现在各种类型的非结构化数据(如办公文档、文本、音频、图片、用户行为等字段长度可变,每个字段又可以由可重复或者不可重复的子字段组成)与半结构化数据,这些数据存储到关系型数据库用于分析会花费更多的时间和金钱。

例如，如果使用 MySQL 存储淘宝的用户行为数据，每天产生的数据量有上百亿甚至更多字节，以 MySQL 每个表存储 4000 万行数据计算，这样每天产生的数据需要 MySQL 分表 100 个以上，可以想象这对机器、运维以及开发是多高的成本。

Google 内部使用的 GFS 以及 MapReduce 技术正好解决了大量数据存储以及分析的问题，而开源社区基于 Google 描述的基于商业硬件集群构建分布式、可扩展的存储和处理系统思想，实现了开源项目 Hadoop 的两个模块，即 HDFS 和 MapReduce。

HDFS 非常适合存储任意的非结构化或者半结构化的数据，它的分布式架构以及简单的扩展方式，使其可以对数据存储提供"无限"的支持，同时因为其可以部署在普通的商业机器集群上，使得存储成本更为低廉，而 MapReduce 则可以帮助用户在需要的时候利用集群中每台机器的能力"分而治之"，恰当地分析这些数据，提供了处理海量数据的核心能力。

GFS 以及 MapReduce 虽然提供了大量数据的存储和数据的分析处理能力，但是对于实时数据的随机存取却无能为力，而且 GFS 适合存储少量的大文件而不适合存储大量的小文件，因为大量的小文件最终会导致元数据的膨胀而可能无法放入主节点的内存。经过多年耕耘，Google 终于在 2006 年发表了一篇"Bigtable: A Distributed Storage System for Structured Data"的论文，这篇论文就是 HBase 的起源，HBase 实现了 BigTable 的架构，如压缩算法、内存操作和布隆过滤器等，HBase 弥补了 Hadoop 的上述缺陷，使得 Hadoop 生态系统更加完善。目前 HBase 在各大公司基本都有使用，如 Facebook 的消息平台、小米的云服务、阿里的 TLog 等许多服务组件。

1.2 NoSQL 与传统 RDBMS

NoSQL，是 Not only SQL 的缩写，泛指非关系型的数据库，如 HBase、MangoDB、Cassandra 等。与传统的关系型数据库（RDBMS）相比，NoSQL 最大的特征就是数据存储不需要一个特定的模式并具有强大的水平扩展能力。

随着大数据时代的到来，特别是超大规模与高并发的社交类型网站的诞生，传统的关系型数据库已经无法满足这些应用系统架构在横向扩展方面的需求了。

- 并发性：关系型数据库更依赖单机性能，对于上万的 QPS，硬盘 IO 无法支撑，并且关系型数据库除了写数据外还要写索引，而 NoSQL，以 HBase 为例，写入为顺序写入，如果能够容忍部分数据的丢失不写 WAL，HBase 写入只需写内存，速度大大提高。同时 HBase 通过分片可以将一台机器的压力均衡地转移到集群的每一台机器。
- 可扩展性：现在一般的在线系统可用性至少都要求达到 99.9%，对于淘宝天猫这种与金钱密切相关的系统，可用性要求就更高了。传统关系型数据库的升级和扩展对于系统的可用性是一个很头疼的问题，可能需要停机迁移数据，重启加载新配置等而且需要运维人员、开发人员、DBA 的密切配合，而 HBase 集群具有线性伸缩、自动容灾和负载均衡的优势，可以很容易地增加或者替换集群节点以扩展集群的存储和计算能力。
- 数据模型：关系型数据库需要为存储的数据预先定义表结构与字段名，而 NoSQL 无须事先为需要存储的数据定义一个模式，这样可以更容易、更灵活去适配各种类型的非结构化数据。

当然，并不是说 NoSQL 已经全面优于关系型数据库了，NoSQL 相比关系型数据库也有很多缺点，例如 HBase 不支持多行事务；基于 LSM 存储模型导致需要读取多个文件来找到需要的数据，这样会牺牲一些读的性能。NoSQL 可以说是对关系型数据库的一种补充。

1.3 应用场景

当互联网概念火热的时候，很多大公司都吹嘘自己是一家互联网公司，现在，很多公司都宣称自己是一家大数据公司。确实现在很多公司都注重数据的采集，而这些数据所使用的存储系统，据我所知，国内各大互联网公司以及知名企业内部基本都使用 HBase。使用 HBase 的业务也非常多，如订单、搜索、用户画像、推荐、监控、安全风控等。HBase 集群在 Facebook 已经达到上万节点的规模，下面列举几个应用场景。

1.3.1 Facebook 用户交互数据

这是一个典型的例子，被引用的次数不计其数，Facebook 的 Like 按钮被点了多少次、有多少人浏览过某个文章、有多少人喜欢这篇文章等数据是由 HBase 的计数器来存储的，发布者能够实时地看到多少人给他点赞、多少人喜欢他的文章。

HBase 在 Facebook 的应用非常成功，也输出了一些著名的技术分享，如 Apache Hadoop Goes Realtime at Facebook 和 Facebook's New Real-Time Messaging System: HBase To Store 135+ Billion Messages A Month。

1.3.2 淘宝 TLog 等

淘宝 TLog 是一个分布式的、可靠的、对大量数据进行收集、分析和展现的系统。主要应用场景是收集大量的运行时日志，然后分析存储，最后提供数据查询和展现。淘宝赫赫有名的鹰眼系统（对请求从开始到结束整个生命周期的追踪，包括哪一步到了哪台机器、每一步花了多长时间、与多少系统有交互等）就是 TLog 的接入方，每天有上万台机器接入 TLog，数据量多达上百 TB，其底层就是使用 HBase 作为存储层。

TLog 的设计也参照了业界著名的 openTSDB 系统——"OpenTSDB is a distributed, scalable Time Series Database (TSDB) written on top of HBase"。感兴趣的读者可以去搜索 OpenTSDB 学习一下。

1.3.3 小米云服务

小米的云服务基本是基于 HBase 存储的。这里以云端数据的同步备份功能为例，用户的照片、联系人、短信、通话记录、米聊等数据中的大部分是非结构化的，重点是数据以用户为隔离，用户只需要访问与自己相关的数据，非常契合 HBase 的分片自动负载均衡，只需使用用户 ID 来进行分区（region，或者说分片），用户量或者数据量的持续增长就可以非常容易地通过为 HBase 集群添加节点来解决。

小米对 HBase 也做了很多贡献，自研的 HBase 自动化部署与监控系统 Minos 已经开源，读者可以去 GitHub 搜索项目 Minos 了解详情。

1.3.4 用户行为数据存储

现在大部分公司都非常注重用户数据的收集（基本上手机上使用的每个 App 都有各种埋点，每一次浏览、点击事件等都会被上传到服务端存储），尤其是用户行为数据的收集，这些数据价值很高，可以用来做很多事情，例如用户画像，既可以为用户构建用户基本信息、行为特征、社交、购买力等静态标签，也可以建立短期的动态标签，例如假设用户刚买车，可能就需要购买一些相关的车载用具。

通过给用户构建一个立体的画像，系统可以近乎实时地分析用户的行为，了解用户的需求，从而实现精准化营销，这对企业尤其是电商企业至关重要。本书我们将会以电商系统的用户行为数据收集作为实战案例来一步一步学习 HBase 相关的知识。

第 2 章

HBase 安装

在生产环境中 HBase 通常与 Hadoop 一起部署分布式集群，为了方便入门体验或者测试，HBase 也提供了单机版的部署包，本章将分别介绍 HBase 的单机版部署与分布式部署。

2.1 单机部署

单机部署可以用来体验或者测试，在 Windows、Linux 以及 Mac 操作系统下都提供了启动脚本，均可一键启动 HBase。

2.1.1 前置条件

单机版的 HBase 只需一个 Java 的运行环境即可启动，表 2-1 描述了各版本 HBase 对 JDK 版本的要求。

表 2-1　Hbase 与 JDK 版本兼容关系

HBase 版本	JDK7.0	JDK8.0
HBase 2.0	不支持	支持
HBase 1.3	支持	支持
HBase 1.2	支持	支持
HBase 1.1	支持	支持但没经过严格测试

2.1.2 下载 HBase

打开 Apache 官网 HBase 下载页面 http://www.apache.org/dyn/closer.cgi/hbase/。

选中一个下载镜像，进入 HBase 版本选择页面，因为企业生产环境一般都使用当前稳定版的 HBase，所以单击 stable 文件夹选择一个稳定版本的 HBase 下载，当前稳定版是 HBase 1.2.6。注意选择.tar.gz 结尾的下载包，src.tar.gz 结尾的下载包是 HBase 的源码。

下载完成后将文件解压，假设我们使用 Windows 系统，将文件解压到 D 盘的 hbase-1.2.6 目录，此时我们可以定义一个环境变量 HBASE_HOME = D:\hbase-1.2.6。

2.1.3 配置 HBase

单机版 HBase 需要配置的唯一文件是%HBASE_HOME%/hbase-1.2.6/conf/hbase-site.xml，需要配置的内容也很少，只需要指定 HBase 和 ZooKeeper 写入数据的路径即可。默认的情况下 HBase 会自动创建一个/tmp 的目录，然后将数据写入这个目录，但是很多系统在重启的时候会删除/tmp 目录，因此通常需要在配置文件中指定数据写入的目录，如代码清单 2-1 所示。

代码清单 2-1　hbase-site.xml

```
<configuration>
  <property>
    <name>hbase.rootdir</name>
    <value>file:///home/mt/hbase</value>
  </property>
  <property>
    <name>hbase.zookeeper.property.dataDir</name>
    <value>/home/mt/zookeeper</value>
  </property>
</configuration>
```

- hbase.rootdir：表示 HBase 数据存储的目录，这里 file:/前缀表示目录是一个本地文件系统，生产环境中一般将 HBase 文件存储构建在 Hadoop 文件系统（HDFS）上，那么该配置项的值类似于 hdfs://mt.com:8020/hbase。
- hbase.zookeeper.property.dataDir：表示 ZooKeeper 数据存储的目录。

2.1.4 启动 HBase

%HBASE_HOME%/hbase-1.2.6/bin/start-hbase.cmd 提供了一个一键启动 HBase 的脚本，

直接运行该 bat 脚本即会帮我们启动整个集群需要的服务，包括 HMaster、HRegionServer 和 ZooKeeper，打开 http://localhost:16010 即可看到 HBase 自带的集群监控页面，如图 2-1 所示，在 Region Servers 下面可以看到当前机器名，这说明 HBase 已经成功启动了。

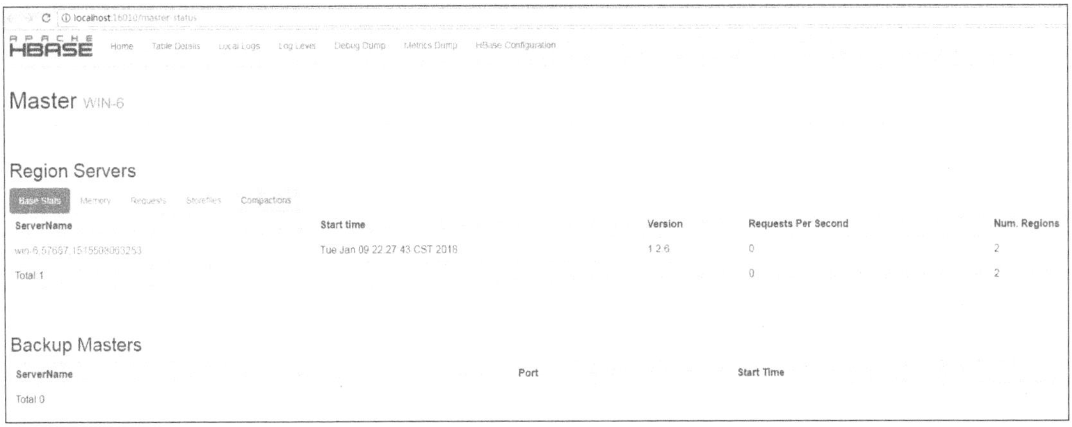

图 2-1　HBase 集群监控页面

> **注意**　如果运行 start-hbase.cmd 时报如下错误：
>
> ```
> 2018-01-09 22:52:57,046 FATAL [WIN-6:61470.activeMasterManager] master.HMaster:
> Failed to become active master
> java.lang.NullPointerException
> ```
>
> 通常是因为缺少 Hadoop 在 Windows 下依赖的组件 winutil.exe 与 hadoop.dll，此时可以从链接 https://github.com/steveloughran/winutil 下载一个最新版本的组件包，当前最新的包是 https://github.com/steveloughran/winutils/tree/master/hadoop-3.0.0/bin。将该目录下面的文件打包全部下载，假设下载后目录为 D:/hadoop-3.0.0/bin，然后添加 Windows 环境变量 HADOOP_HOME=D:/hadoop-3.0.0，再将 %HADOOP_HOME%/bin 添加到 PATH 即可解决。

2.1.5　HBase 初体验

%HBASE_HOME%/hbase-1.2.6/bin/hbase 提供了一个命令行客户端，代码清单 2-2 描述了使用 HBase 命令行创建一个 HBase 表，然后分别对表做 Put、Scan、DeleteAll 数据操作，最后删除这个表。

代码清单 2-2　HBase shell

```
D:\hbase-1.2.6\bin>hbase shell
HBase Shell; enter 'help<RETURN>' for list of supported commands.
```

```
Type "exit<RETURN>" to leave the HBase Shell
Version 1.2.6, r67592f3d062743907f8c5ae00dbbe1ae4f69e5af, Tue Oct 25 18:10:20 CDT 2016
hbase(main):001:0> list
TABLE
0 row(s) in 0.1620 seconds
=> []
hbase(main):002:0> create 's_test', {NAME => 'cf'}
0 row(s) in 2.3110 seconds
=> Hbase::Table - s_test
hbase(main):003:0> put 's_test','rowkey1','cf:v','value1'
0 row(s) in 0.1180 seconds
hbase(main):004:0> scan 's_test'
ROW                      COLUMN+CELL
 rowkey1                 column=cf:v, timestamp=1509634740271, value=value1
1 row(s) in 0.0290 seconds
hbase(main):005:0> deleteall 's_test','rowkye1'
0 row(s) in 0.1170 seconds
hbase(main):006:0> disable 's_test'
0 row(s) in 2.2570 seconds
hbase(main):007:0> drop 's_test'
0 row(s) in 1.2450 seconds
hbase(main):008:0>
```

2.2 分布式部署

在现实生产环境中，为了满足业务的可扩展性与性能要求，HBase 通常会基于 Apache Hadoop 与 Apache ZooKeeper 分布式集群安装。在分布式环境中，一个集群少则数十个节点，多则成百上千个节点，每个节点上通常会同时运行 Hadoop 数据节点（DataNode）、HBase 分区服务器（HRegionServer）进程，此外，还需要有节点运行 HBase HMaster 和 ZooKeeper 进程。在完全分布式的环境中，HBase 集群最少需要 3 个节点（因为 ZooKeeper 选举需要单数个节点，所以 3 个节点中的每个节点运行一个 ZooKeeper 进程）。

表 2-2 描述了 HBase 与 Hadoop 的版本兼容关系。从表 2-2 可以看出，并不是 Hadoop 版本越高，对 HBase 的支持越好。而 Hadoop 2.6.1+与 Hadoop 2.7.1+对 HBase 1.2.X 及以上版本都兼容，因此企业生产环境中一般都采用这两个版本的 Hadoop，而 HBase 则可选择当前最新稳定版 HBase 1.2.6。

表 2-2 HBase 与 Hadoop 版本兼容关系

Hadoop 版本	HBase 1.1.X	HBase 1.2.X	HBase 1.3.X	HBase 2.0.X
Hadoop 2.2.0	未经测试	不支持	不支持	不支持
Hadoop 2.3.X	未经测试	不支持	不支持	不支持
Hadoop 2.4.X	支持	支持	支持	不支持

续表

Hadoop 版本	HBase 1.1.X	HBase 1.2.X	HBase 1.3.X	HBase 2.0.X
Hadoop 2.5.X	支持	支持	支持	不支持
Hadoop 2.6.0	不支持	不支持	不支持	不支持
Hadoop 2.6.1+	未经测试	支持	支持	支持
Hadoop 2.7.0	不支持	不支持	不支持	不支持
Hadoop 2.7.1+	未经测试	支持	支持	支持
Hadoop 2.8.0	不支持	不支持	不支持	不支持
Hadoop 2.8.1	不支持	不支持	不支持	不支持
Hadoop 3.0.0-alphax	未经测试	未经测试	未经测试	未经测试

下面以 3 个节点的集群为例来部署 HBase，3 个节点分别命名为 master1、master2 和 slave1，通常情况下也会将机器名修改为对应的名字以便区分。表 2-3 描述了这 3 个节点上分别将会运行的相关进程。

表 2-3　HBase 集群进程表

进程	master1	master2	slave1
Hadoop 进程	NameNode DataNode DFSZKFailoverController JournalNode	NameNode DataNode DFSZKFailoverController JournalNode	- DataNode - JournalNode
HBase 进程	HMaster HRegionServer	HMaster HRegionServer	- HRegionServer
ZooKeeper 进程	QuorumPeerMain	QuorumPeerMain	QuorumPeerMain

2.2.1　环境准备

下面的步骤描述了在分布式部署之前在操作系统层面需要做的环境准备工作，生产环境操作系统大部分都是 Linux，因此本节也以 Linux 为例。

1. 创建用户

创建用户用来运行所有与 HBase 集群相关的进程，以免权限错乱，代码清单 2-3 描述了在 Linux 环境下如何创建一个 `hadoop` 群组下的 `hadoop` 用户。

代码清单 2-3　创建用户

```
[root@master ~]$ groupadd hadoop
[root@master ~]$ adduser -g hadoop -d /home/hadoop hadoop
```

```
[root@master ~]$ passwd hadoop
[root@master ~]$ mkdir /home/hadoop
[root@master1 home]# sudo chown hadoop:hadoop /home/hadoop
```

2. 修改 hosts 文件

将集群中所有节点的 IP 与名称映射添加到每台机器的/etc/hosts 文件中，这样在之后的各配置项中可以使用节点名称，提高了配置文件的可读性，代码清单 2-4 描述了文件/etc/hosts 添加 IP 与名称映射之后的内容（当集群中节点数量增多时，维护每个节点上的 hosts 文件工作量巨大，此时可以使用域名解析）。

代码清单 2-4　/etc/hosts

```
192.168.0.1 master1
192.168.0.2 master2
192.168.0.3 slave1
```

3. 配置 SSH 免密

配置 SSH 免密可以方便 Hadoop 和 HBase 相关安装文件与配置文件在集群各节点之间传输，以及免密登录。

代码清单 2-5 描述了如何为前面创建的 hadoop 用户生成密钥。

代码清单 2-5　/etc/hosts

```
[root@master ~]$ su - hadoop
[hadoop@master ~]$ ssh-keygen -t rsa
[hadoop@master ~]$ cat ~/.ssh/id_rsa.pub >> ~/.ssh/authorized_keys
[hadoop@master ~]$ chmod 600 ~/.ssh/authorized_keys
```

生成密钥之后需要修改 SSH 配置文件以开启密钥认证，代码清单 2-6 描述了 SSH 配置文件/etc/ssh/sshd_config 需要增加的内容。

代码清单 2-6　/etc/ssh/sshd_config

```
RSAAuthentication yes
PubkeyAuthentication yes
AuthorizedKeysFile      .ssh/authorized_keys
```

最后通过命令[root@master ~]$service sshd restart 重启 SSH 服务即可使免密登录生效。

4. 修改用户可打开文件数与进程数

因为 HBase 存储基于 Apache Hadoop 的 HDFS，当存储数据较大，读取数据时可能需要打开大量的文件，而很多 Linux 操作系统会限制每个用户能够打开的最大文件数（如 centOS 最多打开 1024 个文件）以及最大线程数。通过命令 ulimit -a 查看系统对当前用户限定的最大文件数与线程数，图 2-2 给出的是一台阿里云的 centOS 机器的输出结果。

最大可打开文件数（open files）为 65535：表示该用户最大可打开 65535 个文件，建议配置最小为 10240 个。如果当前用户打开的文件个数超过该配置值，则 Hadoop 或者 HBase 区服务器在运行过程中会报如下错误：`java.net.SocketException: Too many open files`。

用户最大线程数（`max user processes`）为 3895：表示该用户最大可启动 3895 个线程，建议配置为 10000 以上。如果用户启动的线程数超过该配置值，则会报如下错误：`java.lang.OutOfMemoryError:unable to create new native thread`。

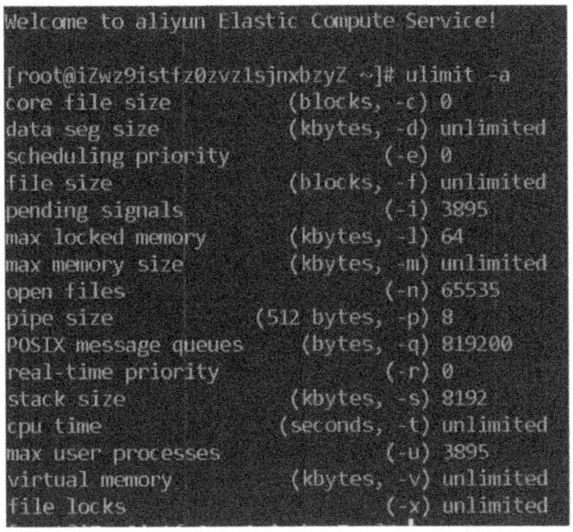

图 2-2　centOS 用户最大文件数与进程数限制

关于如何修改这两个参数，不同操作系统的方法不一样。例如，阿里云机器操作系统是 centOS 5，最大可打开文件数参数见文件/etc/security/limits.conf，只需修改如代码清单 2-7 所示的两行。

代码清单 2-7　/etc/security/limits.conf

```
* soft nofile 65535
* hard nofile 65535
```

2.2.2　ZooKeeper 安装

ZooKeeper 为分布式系统提供配置服务、同步服务、集群管理和选举等功能。HBase 依赖 ZooKeeper 实现主备切换、系统容灾、分区管理和任务管理。ZooKeeper 同样也需要分布式集群部署，如果只部署一个 ZooKeeper 节点，则 ZooKeeper 无法做到容灾高可用；如果 ZooKeeper 节点过多，则会影响 ZooKeeper 同步的效率，因此一般部署 3 个 ZooKeeper 节点，这样既可做到容灾高可用、有效选举，同时也不会带来太大的同步开销。下面以当前稳定版 ZooKeeper-3.4.10 为例在 HBase 集群中部署 ZooKeeper。

（1）下载 ZooKeeper。打开链接 http://apache.website-solution.net/zookeeper/stable/，下载并解压压缩包，假设解压地址为/home/hadoop/zookeeper-3.4.10。

（2）修改 ZooKeeper。配置 ZooKeeper 配置文件路径为/home/hadoop/zookeeper-3.4.10/conf/zoo.cfg，代码清单 2-8 描述了 3 个节点（master1、master2 和 slave1）组成集群的配置。

代码清单 2-8　/home/hadoop/zookeeper-3.4.10/conf/zoo.cfg

```
tickTime=2000
initLimit=10
syncLimit=5
```

```
dataDir=/home/hadoop/zookeeper-3.4.10/data
dataLogDir=/home/hadoop/zookeeper-3.4.10/datalog
clientPort=2181
# 开启日志与镜像（ZooKeeper 快照）文件自动清理，否则运行一段时间后日志与镜像文件会占满磁盘
autopurge.purgeInterval=1
#以下是集群节点配置，格式为 "机器名.myid:数据通信端口:ZooKeeper Leader 选举端口"
server.1=master1:2888:3888
server.2=master2:2888:3888
server.3=slave1:2888:3888
```

（3）创建 myid 身份标识文件。每个节点都需要创建一个 myid 文件来用作这个节点的身份标识。myid 父文件夹路径为第 2 步中配置的 dataDir 目录，文件内容为一个数字，即 zoo.cfg 中集群节点配置项 server.x 的 x，因此这里 3 台机器的 myid 文件内容分别为 1、2 和 3。

2.2.3　Hadoop 安装

HBase 存储基于 Hadoop 的 HDFS（Hadoop Distributed File System），内置的一些统计任务也需依赖 MapReduce。以下步骤描述如何修改 Hadoop 各配置文件来搭建 3 台机器（master1、master2 和 slave1）的 Hadoop 集群。

（1）下载 Hadoop。打开链接 http://hadoop.apache.org/releases.html，下载并解压压缩包，这里下载对 HBase 兼容性较好的 Hadoop 版本 Hadoop 2.6.5，解压到目录/home/hadoop/hadoop-2.6.5。

（2）修改配置文件。需要修改的配置文件比较多，这些配置文件都位于/home/hadoop/hadoop-2.6.5/etc/hadoop。本节列出的配置文件包含了 Hadoop 运行时所需的一些重要配置项。

代码清单 2-9 描述了与 Hadoop 运行环境相关的配置文件 hadoop-env.sh，读者可以重点关注一下 JVM 参数的配置。

代码清单 2-9　/home/hadoop/hadoop-2.6.5/etc/hadoop/hadoop-env.sh

```
#Java 安装目录，建议使用 JDK 1.8：
export JAVA_HOME=<<JDK install path>>
#配置 Hadoop NameNode 运行堆内存为 12 GB
export HADOOP_NAMENODE_OPTS="-Xmx12g -Xms12g
-Dhadoop.security.logger=${HADOOP_SECURITY_LOGGER:-INFO,RFAS}
-Dhdfs.audit.logger=${HDFS_AUDIT_LOGGER:-INFO,NullAppender} $HADOOP_NAMENODE_OPTS"
#配置 Hadoop DataNode 运行堆内存为 12 GB
export HADOOP_DATANODE_OPTS="-Xmx12g -Xms12g -Dhadoop.security.logger=ERROR,RFAS
$HADOOP_DATANODE_OPTS"
#配置 Hadoop 进程 id 文件路径，默认路径为/tmp 操作系统重启的时候可能会被清除
export HADOOP_PID_DIR=/home/hadoop/hadoop-2.6.5
```

代码清单 2-10 描述了 Hadoop 的核心配置文件 core-site.xml。

代码清单 2-10　/home/hadoop/hadoop-2.6.5/etc/hadoop/core-site.xml

```xml
<?xml version="1.0" encoding="UTF-8"?>
<?xml-stylesheet type="text/xsl" href="configuration.xsl"?>
<configuration>
  <property>
    <name>fs.defaultFS</name>
    <value>hdfs://mtcluster</value>
    <final>true</final>
  </property>

  <property>
    <name>io.file.buffer.size</name>
    <value>131072</value>
  </property>

  <property>
    <name>hadoop.tmp.dir</name>
    <value>/home/hadoop/tmp/${user.name}</value>
  </property>

  <!--ZooKeeper集群的连接地址-->
  <property>
    <name>ha.zookeeper.quorum</name>
    <value>master1:2181,master2:2181,slave1:2181</value>
  </property>

  <!--数据压缩编码-->
  <property>
    <name>io.compression.codecs</name>
    <value>
      org.apache.hadoop.io.compress.GzipCodec,
      org.apache.hadoop.io.compress.DefaultCodec,
      org.apache.hadoop.io.compress.SnappyCodec
    </value>
  </property>
  <property>
    <name>fs.trash.interval</name>
    <value>1440</value>
  </property>

  <property>
    <name>fs.trash.checkpoint.interval</name>
    <value>720</value>
  </property>

  <property>
     <name>ha.failover-controller.cli-check.rpc-timeout.ms</name>
     <value>60000</value>
  </property>
```

```xml
    <property>
        <name>ipc.client.connect.timeout</name>
        <value>60000</value>
    </property>
    <!--当需要开启机柜感知的情况下使用,指定了机柜感知脚本路径,具体机柜感知如何开启可见 9.2.3 节-->
    <property>
        <name>topology.script.file.name</name>
        <value>/home/hadoop/hadoop-2.6.5/etc/hadoop/topology.sh</value>
    </property>
</configuration>
```

代码清单 2-11 描述了配置文件 hdfs-site.xml。

代码清单 2-11　/home/hadoop/hadoop-2.6.5/etc/hadoop/hdfs-site.sh

```xml
<?xml version="1.0" encoding="UTF-8"?>
<?xml-stylesheet type="text/xsl" href="configuration.xsl"?>
<configuration>
    <property>
        <name>dfs.nameservices</name>
        <value>mtcluster</value>
    </property>

    <property>
        <name>dfs.ha.namenodes.mtcluster</name>
        <value>nn1,nn2</value>
    </property>

    <property>
        <name>dfs.namenode.rpc-address.mtcluster.nn1</name>
        <value>master1:9000</value>
    </property>

    <property>
        <name>dfs.namenode.rpc-address.mtcluster.nn2</name>
        <value>master2:9000</value>
    </property>

    <property>
        <name>dfs.namenode.http-address.mtcluster.nn1</name>
        <value>master1:50070</value>
    </property>

    <property>
        <name>dfs.namenode.http-address.mtcluster.nn2</name>
        <value>master2:50070</value>
    </property>

    <property>
        <name>dfs.namenode.servicerpc-address.mtcluster.nn1</name>
```

```xml
    <value>master1:53310</value>
</property>

<property>
    <name>dfs.namenode.servicerpc-address.mtcluster.nn2</name>
    <value>master2:53310</value>
</property>

<property>
    <name>dfs.namenode.name.dir.mtcluster</name>
    <value>/home/hadoop/data01/nn</value>
    <final>true</final>
</property>

<property>
    <name>dfs.namenode.shared.edits.dir</name>
    <value>qjournal://master1:8485;master2:8485;slave1:8485/mtcluster</value>
</property>

<!--为了避免 Hadoop NameNode 单点故障，一般会在集群中部署两个 NameNode，该参数设置为 true
    的时候表示开启高可用，即当当前活跃的 NameNode 故障时，另外一个 NameNode 会自动从备用状态转为
    活跃状态提供服务-->
<property>
    <name>dfs.ha.automatic-failover.enabled</name>
    <value>true</value>
</property>

<property>
    <name>dfs.client.failover.proxy.provider.mtcluster</name>
    <value>
    org.apache.hadoop.hdfs.server.namenode.ha.ConfiguredFailoverProxyProvider
    </value>
</property>

<property>
     <name>dfs.journalnode.edits.dir</name>
     <value>/home/hadoop/data01/tmp/journal</value>
</property>

<property>
     <name>dfs.ha.fencing.methods</name>
     <value>sshfence(hadoop:16120)</value>
</property>

<property>
     <name>dfs.ha.fencing.ssh.private-key-files</name>
     <value>/home/hadoop/.ssh/id_rsa</value>
</property>

<!--指定 HDFS DataNode 数据存储的路径，一般建议一台机器挂载多个盘，一方面可以增大存储容量，另
    一方面可以减少磁盘单点故障以及磁盘读写压力-->
```

```xml
<property>
  <name>dfs.datanode.data.dir</name>
  <value>
    /home/hadoop/data01/dn,/home/hadoop/data02/dn,/home/hadoop/data03/dn,/home/hadoop/data04/dn
  </value>
  <final>true</final>
</property>

<property>
  <name>dfs.namenode.checkpoint.dir.mtcluster</name>
  <value>/home/hadoop/data01/dfs/namesecondary</value>
  <final>true</final>
</property>

<!-- 表示每个 DataNode 上面需要预留的空间，以免 DataNode 写数据将磁盘耗尽 -->
<property>
  <!-- space amount in bytes reserved on the storage volumes for non-HDFS use -->
  <name>dfs.datanode.du.reserved</name>
  <value>10000000000</value>
  <final>true</final>
</property>

<property>
  <name>dfs.hosts.exclude</name>
  <value>/home/hadoop/hadoop-2.6.5/etc/hadoop/hosts.exclude</value>
</property>

<property>
  <name>dfs.hosts</name>
  <value>/home/hadoop/hadoop-2.6.5/etc/hadoop/hosts.allow</value>
</property>

<!--每个数据块保留的备份数据，用来保证数据的高可用 -->
<property>
  <name>dfs.replication</name>
  <value>3</value>
  <final>true</final>
</property>

<!-- 限制 HDFS 负载均衡运行时占用的最大带宽 -->
<property>
   <name>dfs.datanode.balance.bandwidthPerSec</name>
   <value>104857600</value>
   <description>
       Specifies the maximum amount of bandwidth that each datanode can utilize
       for the balancing purpose in term of the number of bytes per second.
   </description>
</property>
```

```xml
<!-- 设置DataNode可以容忍的坏盘数量，如果配置为0，则表示不容许有坏盘。当有坏盘时，DataNode启
     动会失败-->
<property>
    <name>dfs.datanode.failed.volumes.tolerated</name>
    <value>1</value>
</property>

<!--配合HBase或者其他dfs客户端使用，表示开启短路径读，可以用来优化客户端性能，需要配合dfs.
    domain.socket.path使用，具体可参见9.2.6节-->
 <property>
    <name>dfs.client.read.shortcircuit</name>
    <value>true</value>
 </property>
 <property>
    <name>dfs.domain.socket.path</name>
    <value>/home/hadoop/hadoop-2.6.5/dn_socket</value>
 </property>
</configuration>
```

代码清单 2-12 描述了与 MapReduce 任务相关的配置文件 mapred-site.xml。

代码清单 2-12　/home/hadoop/hadoop-2.6.5/etc/hadoop/mapred-site.xml

```xml
<?xml version="1.0"?>
<?xml-stylesheet type="text/xsl" href="configuration.xsl"?>
<configuration>
  <property>
    <name>mapreduce.framework.name</name>
    <value>yarn</value>
  </property>

  <property>
    <name>mapreduce.jobhistory.address</name>
    <value>master1:10020</value>
  </property>

  <property>
    <name>mapreduce.jobhistory.webapp.address</name>
    <value>master1:19888</value>
  </property>

  <property>
  <name>mapreduce.cluster.local.dir</name>
    <value>
      /home/hadoop/data01/mapred/local
    </value>
  </property>

  <property>
```

```xml
    <name>mapreduce.jobtracker.system.dir</name>
    <value>${hadoop.tmp.dir}/mapred/system</value>
</property>

<property>
    <name>mapreduce.jobtracker.heartbeats.in.second</name>
    <value>100</value>
</property>

<property>
    <name>mapreduce.tasktracker.outofband.heartbeat</name>
    <value>true</value>
</property>

<property>
    <name>mapreduce.jobtracker.staging.root.dir</name>
    <value>${hadoop.tmp.dir}/mapred/staging</value>
</property>


<property>
    <name>mapreduce.cluster.temp.dir</name>
    <value>${hadoop.tmp.dir}/mapred/temp</value>
</property>

<property>
    <name>yarn.app.mapreduce.am.resource.mb</name>
    <value>2560</value>
</property>

<property>
    <name>yarn.app.mapreduce.am.command-opts</name>
    <value>-Xmx2048m</value>
</property>

<property>
    <name>mapreduce.map.memory.mb</name>
    <value>2560</value>
</property>

<property>
    <name>mapreduce.map.java.opts</name>
    <value>-Xmx2048m</value>
</property>

<property>
    <name>mapreduce.reduce.memory.mb</name>
    <value>2560</value>
</property>
```

```xml
<property>
  <name>mapreduce.reduce.java.opts</name>
  <value>-Xmx2048m</value>
</property>

<property>
  <name>mapreduce.task.io.sort.mb</name>
  <value>1024</value>
</property>

<property>
  <name>mapreduce.map.cpu.vcores</name>
  <value>1</value>
</property>

<property>
  <name>mapreduce.reduce.cpu.vcores</name>
  <value>1</value>
</property>

<property>
  <name>mapreduce.map.output.compress</name>
  <value>true</value>
</property>

<property>
  <name>mapreduce.map.output.compress.codec</name>
  <value>com.hadoop.compression.lzo.LzoCodec</value>
</property>

<property>
  <name>mapreduce.output.fileoutputformat.compress</name>
  <value>true</value>
</property>

<property>
  <name>mapreduce.output.fileoutputformat.compress.type</name>
  <value>BLOCK</value>
</property>

<property>
  <name>mapreduce.output.fileoutputformat.compress.codec</name>
  <value>com.hadoop.compression.lzo.LzoCodec</value>
</property>

<property>
  <name>mapreduce.task.timeout</name>
  <value>180000</value>
</property>
```

```xml
<property>
    <name>mapreduce.jobtracker.handler.count</name>
    <value>60</value>
</property>

<property>
    <name>mapreduce.reduce.shuffle.parallelcopies</name>
    <value>20</value>
</property>

<property>
    <name>mapreduce.tasktracker.http.threads</name>
    <value>60</value>
</property>

<property>
    <name>mapreduce.job.reduce.slowstart.completedmaps</name>
    <value>0.8</value>
</property>

<property>
    <name>mapreduce.task.io.sort.factor</name>
    <value>60</value>
</property>

<property>
    <name>mapreduce.reduce.input.buffer.percent</name>
    <value>0.8</value>
</property>

<property>
    <name>mapreduce.map.combine.minspills</name>
    <value>3</value>
</property>

<property>
      <name>mapreduce.jobtracker.taskscheduler</name>
      <value>org.apache.hadoop.mapred.FairScheduler</value>
</property>

</configuration>
```

代码清单 2-13 描述了配置文件 yard-site.xml,该文件同样定义了与 MapReduce 任务运行相关的配置。

代码清单 2-13   /home/hadoop/hadoop-2.6.5/etc/hadoop/yard-site.xml

```xml
<?xml version="1.0"?>
<configuration>
  <property>
```

```xml
    <name>yarn.resourcemanager.hostname</name>
    <value>master1</value>
</property>
<property>
    <name>yarn.nodemanager.aux-services</name>
    <value>mapreduce_shuffle,spark_shuffle</value>
</property>
<property>
    <name>yarn.nodemanager.aux-services.spark_shuffle.class</name>
    <value>org.apache.spark.network.yarn.YarnShuffleService</value>
</property>
<property>
    <name>yarn.nodemanager.aux-services.mapreduce_shuffle.class</name>
    <value>org.apache.hadoop.mapred.ShuffleHandler</value>
</property>
<property>
    <name>yarn.resourcemanager.scheduler.class</name>
    <value>org.apache.hadoop.yarn.server.resourcemanager.scheduler.capacity.CapacityScheduler</value>
</property>
<property>
    <name>yarn.resourcemanager.scheduler.monitor.enable</name>
    <value>true</value>
</property>
<property>
    <name>yarn.resourcemanager.scheduler.monitor.policies</name>
    <value>org.apache.hadoop.yarn.server.resourcemanager.monitor.capacity.ProportionalCapacityPreemptionPolicy</value>
</property>
<property>
    <name>yarn.resourcemanager.monitor.capacity.preemption.observe_only</name>
    <value>true</value>
</property>
<property>
    <name>yarn.resourcemanager.monitor.capacity.preemption.total_preemption_per_round</name>
    <value>1</value>
</property>
<property>
<name>yarn.resourcemanager.monitor.capacity.preemption.max_ignored_over_capacity</name>
    <value>0</value>
</property>
<property>
    <name>yarn.resourcemanager.monitor.capacity.preemption.natural_termination_factor</name>
    <value>1</value>
</property>
<property>
    <name>yarn.nodemanager.resource.cpu-vcores</name>
```

```xml
    <value>12</value>
  </property>
  <property>
    <name>yarn.nodemanager.local-dirs</name>
    <value>
       /home//hadoop/data01/nm-local-dir
    </value>
    <final>true</final>
  </property>
  <property>
    <name>yarn.log-aggregation.retain-seconds</name>
    <value>604800</value>
  </property>
  <property>
    <name>yarn.application.classpath</name>
    <value>
      $HADOOP_CONF_DIR,
      $HADOOP_COMMON_HOME/share/hadoop/common/*,
      $HADOOP_COMMON_HOME/share/hadoop/common/lib/*,
      $HADOOP_HDFS_HOME/share/hadoop/hdfs/*,
      $HADOOP_HDFS_HOME/share/hadoop/hdfs/lib/*,
      $HADOOP_MAPRED_HOME/share/hadoop/mapreduce/*,
      $HADOOP_MAPRED_HOME/share/hadoop/mapreduce/lib/*,
      $HADOOP_YARN_HOME/share/hadoop/yarn/*,
      $HADOOP_YARN_HOME/share/hadoop/yarn/lib/*
    </value>
  </property>
  <property>
     <name>yarn.log.server.url</name>
     <value>http://master1:19888/jobhistory/logs</value>
  </property>
  <property>
     <description>The hostname of the Timeline service web application.</description>
     <name>yarn.timeline-service.hostname</name>
     <value>master1</value>
  </property>
  <property>
     <description>Address for the Timeline server to start the RPC server.</description>
     <name>yarn.timeline-service.address</name>
     <value>${yarn.timeline-service.hostname}:10200</value>
  </property>
  <property>
     <description>The http address of the Timeline service web application.</description>
     <name>yarn.timeline-service.webapp.address</name>
     <value>${yarn.timeline-service.hostname}:8188</value>
  </property>
  <property>
     <description>The https address of the Timeline service web application.</description>
     <name>yarn.timeline-service.webapp.https.address</name>
     <value>${yarn.timeline-service.hostname}:8190</value>
```

```xml
    </property>
    <property>
        <description>Handler thread count to serve the client RPC requests.</description>
        <name>yarn.timeline-service.handler-thread-count</name>
        <value>60</value>
    </property>
    <property>
        <name>yarn.resourcemanager.resource-tracker.client.thread-count</name>
        <value>60</value>
    </property>
    <property>
        <name>yarn.resourcemanager.scheduler.client.thread-count</name>
        <value>60</value>
    </property>
</configuration>
```

代码清单 2-14 描述了配置文件 slaves，该文件每行代表一个 Hadoop 集群中的数据节点。

代码清单 2-14　/home/hadoop/hadoop-2.6.5/etc/hadoop/slaves

```
master1
master2
slave1
```

> **提示**　当上述配置文件均准备完成后，可以通过 scp 命令将 hadoop-2.6.5 文件夹分发到集群中的每一个节点：
>
> ```
> scp -r hadoop-2.6.5 hadoop@master2:/home/hadoop/
> scp -r hadoop-2.6.5 hadoop@slave1:/home/hadoop/
> ```

至此，Hadoop 相关配置文件已经准备完毕，接下来待 HBase 等安装完成后一起启动 Hadoop 集群。

2.2.4　HBase 安装

以下步骤以 HBase 当前稳定版本 HBase 1.2.6 描述了如何下载安装 HBase。

（1）下载 HBase。打开链接 http://apache.01link.hk/hbase/stable/，下载并解压压缩包，解压到目录/home/hadoop/hbase-1.2.6。

（2）修改配置文件。HBase 配置文件均位于/home/hadoop/hbase-1.2.6/config 目录。由于 HBase 依赖 Hadoop 的存储系统 HDFS，因此首先需要复制 Hadoop 的配置文件 hdfs-site.xml 到 HBase，如代码清单 2-15 所示。

代码清单 2-15　复制 hdfs-site.xml

```
cp /home/hadoop/hadoop-2.6.5/etc/hadoop/hdfs-site.xml /home/hadoop/hbase-1.2.6/conf/
```

代码清单 2-16 描述了与 HBase 运行环境相关的配置脚本文件 hbase-env.sh。

代码清单 2-16　/home/hadoop/hbase-1.2.6/config/hbase-env.sh

```
#Java 安装目录，建议使用 JDK1.8：
export JAVA_HOME=<<JDK install path>>
export HADOOP_HOME=/home/hadoop/hadoop-2.6.5
export HBASE_HOME=/home/hadoop/hbase-1.2.6
export HADOOP_CONF_DIR=${HADOOP_HOME}/etc/hadoop
#配置 HBase HMaster 进程运行堆内存为 16GB
export HBASE_MASTER_OPTS="-Xmx16384m"
#配置 HBase 分区服务器进程运行 JVM 参数，读者需要根据节点内存做调整，关于 JVM 调优可见第 9 章
export HBASE_REGIONSERVER_OPTS="-Xss256k -Xmx24g -Xms24g -Xmn2g
-XX:MaxDirectMemorySize=24g -XX:SurvivorRatio=3 -XX:+UseParNewGC
-XX:+UseConcMarkSweepGC -XX:MaxTenuringThreshold=10
-XX:CMSInitiatingOccupancyFraction=80 -XX:+UseCMSCompactAt FullCollection
-XX:+UseCMSInitiatingOccupancyOnly
-XX:+DisableExplicitGC -XX:+He apDumpOnOutOfMemoryError
-verbose:gc -XX:+PrintGCDateStamps -XX:+PrintGCTimeStamps -XX:+PrintGCDetails
-XX:+PrintTenuringDistribution -XX:+PrintCommandLineFlags
-XX:ErrorFile=${HBASE_HOME}/logs/hs_err_pid%p-$(hostname).log
-XX:HeapDumpPath=${HBASE_HOME}/logs/
-Xloggc:${HBASE_HOME}/logs/gc-$(hostname)-hbase.log"
export HBASE_SSH_OPTS="-p 16120"
export HBASE_LOG_DIR=${HBASE_HOME}/logs
export HBASE_PID_DIR=${HBASE_HOME}/pid
export HBASE_MANAGES_ZK=false
#配置 HBase 运行相关的依赖库地址
export LD_LIBRARY_PATH=${LD_LIBRARY_PATH}:${HADOOP_HOME}/lib/native/:/usr/local/lib
export HBASE_LIBRARY_PATH=${HBASE_LIBRARY_PATH}:${HBASE_HOME}/lib/native/Linux-amd64-64:/usr/local/lib/:${HADOOP_HOME}/lib/native/
```

代码清单 2-17 描述了 HBase 的核心配置文件 hbase-site.xml。

代码清单 2-17　/home/hadoop/hbase-1.2.6/config/hbase-site.xml

```xml
<?xml version="1.0"?>
<?xml-stylesheet type="text/xsl" href="configuration.xsl"?>
<configuration>
    <property>
        <name>hbase.rootdir</name>
        <value>hdfs://mtcluster/user/hbase</value>
    </property>
    <property>
        <name>zookeeper.znode.parent</name>
        <value>/hbase</value>
    </property>
    <property>
        <name>hbase.cluster.distributed</name>
        <value>true</value>
    </property>
    <property>
        <name>hbase.tmp.dir</name>
```

```xml
        <value>/home/hadoop/data01/hbase/hbase_tmp</value>
</property>
<property>
        <name>hbase.zookeeper.property.dataDir</name>
        <value>/home/hadoop/data01/hbase/zookeeper_data</value>
</property>
<property>
        <name>hbase.master.port</name>
        <value>61000</value>
</property>
<property>
        <name>hbase.zookeeper.quorum</name>
        <value>master1,master2,slave1</value>
</property>
<property>
        <name>hbase.zookeeper.property.clientPort</name>
        <value>2181</value>
 </property>
<property>
        <name>hbase.client.keyvalue.maxsize</name>
        <value>0</value>
</property>
<property>
        <name>hbase.master.distributed.log.splitting</name>
        <value>true</value>
</property>
 <!--一次 RPC 请求读取的数据行数，该参数设置有助于优化读取效率，详细介绍可见 9.1.2 节 -->
<property>
        <name>hbase.client.scanner.caching</name>
        <value>500</value>
</property>
<property>
        <name>hfile.block.cache.size</name>
        <value>0.2</value>
</property>
 <!--当分区中 StoreFile 大小超过该值时，该分区可能会被拆分（受是否开启了自动 split 影响），
    一般线上集群会关闭自动 split 以免影响性能，因此会将该值设置的比较大，如 100G-->
<property>
        <name>hbase.hregion.max.filesize</name>
        <value>107374182400</value><!-- 100G -->
</property>
<property>
        <name>hbase.hregion.memstore.flush.size</name>
        <value>268435456</value><!-- 256M -->
</property>
<property>
        <name>hbase.regionserver.handler.count</name>
        <value>200</value>
</property>
<property>
        <name>hbase.regionserver.global.memstore.lowerLimit</name>
```

```xml
        <value>0.38</value>
    </property>
    <property>
        <name>hbase.regionserver.global.memstore.size</name>
        <value>0.45</value>
    </property>
    <property>
        <name>hbase.hregion.memstore.block.multiplier</name>
        <value>8</value>
    </property>
    <property>
        <name>hbase.server.thread.wakefrequency</name>
        <value>1000</value>
    </property>
    <property>
        <name>hbase.rpc.timeout</name>
        <value>400000</value>
    </property>
    <!--当HStore的StoreFile数量超过该配置值时，MemStore刷新到磁盘之前需要进行拆分（split）
        或者压缩（compact），除非超过hbase.hstore.blockingWaitTime配置的时间。因此，当禁止
        自动主压缩（major compact）的时候该配置项一定要注意配置一个较大的值-->
    <property>
        <name>hbase.hstore.blockingStoreFiles</name>
        <value>5000</value>
    </property>
    <property>
        <name>hbase.client.scanner.timeout.period</name>
        <value>1000000</value>
    </property>
    <property>
        <name>zookeeper.session.timeout</name>
        <value>180000</value>
    </property>
    <property>
        <name>hbase.regionserver.optionallogflushinterval</name>
        <value>5000</value>
    </property>
    <property>
        <name>hbase.client.write.buffer</name>
        <value>5242880</value>
    </property>
    <!--当HStore的StoreFile数量超过该配置的值时，可能会触发压缩，该值不能设置得过大，否则
        会影响读性能，一般建议设置为3~5-->
    <property>
        <name>hbase.hstore.compactionThreshold</name>
        <value>5</value>
    </property>
    <property>
        <name>hbase.hstore.compaction.max</name>
        <value>12</value>
```

```xml
    </property>
    <!--将该值设置为1以禁止线上表的自动拆分(split),可以在建表的时候预分区或者之后手动分区-->
    <property>
        <name>hbase.regionserver.regionSplitLimit</name>
        <value>1</value>
    </property>
    <property>
        <name>hbase.regionserver.thread.compaction.large</name>
        <value>5</value>
    </property>
    <property>
        <name>hbase.regionserver.thread.compaction.small </name>
        <value>8</value>
    </property>
    <property>
        <name>hbase.master.logcleaner.ttl</name>
        <value>3600000</value>
    </property>
    <property>
        <name>hbase.bucketcache.ioengine</name>
        <value>offheap</value>
    </property>
    <property>
        <name>hbase.bucketcache.percentage.in.combinedcache</name>
        <value>0.9</value>
    </property>
    <property>
        <name>hbase.bucketcache.size</name>
        <value>16384</value>
    </property>
    <property>
        <name>hbase.replication</name>
        <value>true</value>
    </property>
     <!--开启镜像(snapshot)功能支持-->
    <property>
        <name>hbase.snapshot.enabled</name>
        <value>true</value>
    </property>
     <!--复制带宽限制:默认为0表示不限速,如果复制带宽限速100MB,即 100*1024*1024-->
    <property>
         <name>replication.source.per.peer.node.bandwidth</name>
         <value>104857600</value>
    </property>
     <!--复制主集群能够选择从集群的服务器百分比,如果从集群有3台机器,则配置值1表示主集群能够
        选择所有的3台机器用来推送复制数据-->
    <property>
         <name>replication.source.ratio</name>
         <value>1</value>
    </property>
```

```xml
<!--配置主压缩的间隔时间,0表示禁止自动主压缩,如果是线上响应时间敏感的应用,则建议禁止而
    等到非高峰期手动压缩,否则很有可能导致 HBase 响应超时而引起性能抖动-->
<property>
    <name>hbase.hregion.majorcompaction</name>
    <value>0</value>
</property>
</configuration>
```

代码清单 2-18 描述了配置文件 regionservers，该文件每行代表一个 HBase 集群中分区服务器（RegionServer）节点。

代码清单 2-18　/home/hadoop/hbase-1.2.6/config/regionservers
```
master1
master2
slave1
```

为了解决 HBase 主节点单点问题，一般在 HBase 集群中部署两个主节点，代码清单 2-19 描述集群中备用的主节点。

代码清单 2-19　/home/hadoop/hbase-1.2.6/config/backup-masters
```
master2
```

配置文件中一些关键项目都添加了注释，读者可以先行启动集群，然后再跟进需求修改响应配置。

2.2.5　启动集群

HBase 提供了 start-hbase.sh 脚本启动 HBase 集群，为了更好地了解集群中每个节点应该运行什么进程，下面将哪个节点将运行什么命令启动哪个进程一一列出，由软件之间的依赖顺序可以得到启动顺序应该是 ZooKeeper→Hadoop→HBase，停止顺序则相反。

（1）启动 ZooKeeper。在 master1、master2 和 slave1 上执行如下命令启动 ZooKeeper：

/home/hadoop/zookeeper-3.4.6/bin/zkServer.sh start

（2）启动 Hadoop 日志节点。在 master1、master2 和 slave1 上执行如下命令启动 JournalNode：

/home/hadoop/hadoop-2.6.5/sbin/hadoop-daemon.sh start journalnode

（3）格式化 Hadoop 名称节点（NameNode）。在 master1 上执行如下命令格式化名称节点：

/home/hadoop/hadoop-2.6.5/bin/hadoop namenode -format

（4）启动 Hadoop 名称节点。在 master1 上执行如下命令启动名称节点：

/home/hadoop/hadoop-2.6.5/sbin/hadoop-daemon.sh start namenode

为了保证 Hadoop 的高可用，通常会在集群部署两个名称节点，因此还需要在 master2 启动备用的名称节点，启动之前需要将 master1 的 namenode 元数据复制到 master2，否则 master2 也会生成一个集群唯一标识符 clusterId，也就是被当成了另外一个集群的主名称节点。

在 master1 执行如下命令复制元数据：

```
scp -r /home/hadoop/data01/nn hadoop@master2:/home/hadoop/data01
```

在 master2 上执行如下命令启动备用名称节点：

```
/home/hadoop/hadoop-2.6.5/sbin/hadoop-daemon.sh start namenode
```

（5）启动 Hadoop ZKFC。ZKFC 全称是 ZooKeeper failover controller，意为基于 ZooKeeper 的自动容灾控制，用来负责 Hadoop 名称节点的自动容灾。在 master1 上执行如下命令格式化 ZKFC：

```
/home/hadoop/hadoop-2.6.5/sbin/hadoop-daemon.sh start zkfc -formatZK
```

在 master1 和 master2 上执行如下命令启动 ZKFC：

```
/home/hadoop/hadoop-2.6.5/sbin/hadoop-daemon.sh start zkfc
```

（6）启动 Hadoop 数据节点。在 master1、master2 和 slave1 上执行如下命令启动数据节点：

```
/home/hadoop/hadoop-2.6.5/sbin/hadoop-daemon.sh start datanode
```

（7）启动 HBase 主节点 HMaster。在 master1 和 master2 上执行如下命令启动 HMaster 进程：

```
/home/hadoop/hbase-1.2.6/bin/hbase-daemon.sh start master
```

（8）启动 HBase 分区服务器。在 master1、master2 和 slave1 上执行如下命令启动 HRegionServer 进程：

```
/home/hadoop/hbase-1.2.6/bin/hbase-daemon.sh start regionserver
```

2.3 集群增删节点

分布式扩展的弹性伸缩是 HBase 与传统数据库相比最大的优点之一，便于 HBase 集群增加删除节点。

2.3.1 增加节点

下面的步骤描述了如何增加一个新节点到 HBase 集群。假设新增的节点 IP 为

192.168.0.4，机器名沿用之前的命名规则为 slave2。

（1）修改 /etc/hosts 文件。在 master1、master2 和 slave1 上分别运行如下命令以追加新机器名与 IP 绑定到 hosts 文件，然后将 hosts 文件复制到 slave2：

```
echo '192.168.0.4' slave2 >> /etc/hosts
```

（2）修改 Hadoop /home/hadoop/hadoop-2.6.5/etc/hadoop/slaves 文件。在 master1、master2 和 slave1 上分别运行如下命令以追加新机器名到 slaves 文件：

```
echo 'slave2 >> /home/hadoop/hadoop-2.5.6/etc/hadoop/slaves
```

（3）刷新集群节点。在名称节点 master1 上运行如下命令以刷新 Hadoop 集群数据节点列表：

```
/home/hadoop/hadoop-2.5.6/bin/hadoop dfsadmin -refreshNodes
```

（4）启动 Hadoop 数据节点 slave2。在 slave2 上运行如下命令以启动 Hadoop 数据节点：

```
/home/hadoop/hadoop-2.6.5/sbin/hadoop-daemon.sh start datanode
```

（5）运行负载均衡。新加入的节点上无数据，如果新加入节点后立即在新节点上启动 HBase 分区服务器进程 HRegionServer，就会导致新节点需要发起大量网络请求去读取其他节点上的数据，因此一般新加入节点后需要在 Hadoop 集群上运行负载均衡以平衡集群各节点的数据容量。在集群任意节点执行如下命令以运行负载均衡，参数 5 表示如果集群每个节点存储百分比差距在 5% 以内则结束该次负载均衡：

```
/home/hadoop/hadoop-2.5.6/bin/hdfs balancer -threshold 5
```

（6）增加 HBase 分区服务器。在 master1、master2 和 slave1 上运行如下命令以将新节点加入 HBase 分区服务器列表：

```
echo 'slave2 >> /home/hadoop/hbase-1.2.6/conf/regionservers
```

（7）启动 HBase 分区服务器。在 slave2 上运行如下命令以启动 HBase 分区服务器，至此新增节点已经完成：

```
/home/hadoop/hbase-1.2.6/bin/hbase-daemon.sh start regionserver
```

2.3.2 删除节点

Hadoop 与 HBase 的分布式集群特征使其可以对节点故障做到自动容灾与负载转移，但是有时候当集群中某些机器硬件配置过时或者需要滚动升级的时候，也可以很方便地将节点下线，以下步骤描述了如何将刚刚增加的节点 slave 2 下线。

（1）下线 HBase 分区服务器。在 slave2 节点（也可以在集群任意节点）上执行如下命令以优雅的形式下线分区服务器，下线过程中会将该分区服务器负责的所有分区都转移到集群中其他在线的节点：

```
/home/hadoop/hbase-1.2.6/bin/graceful_stop.sh slave2
```

（2）Hadoop 配置数据节点下线。在/home/hadoop/hadoop-2.6.5/etc/hadoop/hdfs-site.xml 中增加如代码清单 2-20 所示配置项，注意到 slaves 文件之前已经存在，只需要新建 exclude-slaves 文件，然后在该文件中添加一行内容 slave2，表示需要将 slave2 节点排除。

代码清单 2-20　/home/hadoop/hadoop-2.6.5/etc/hadoop/hdfs-site.xml

```
<property>
    <name>dfs.hosts</name>
    <value>/home/hadoop/hadoop2.6.5/etc/hadoop/slaves</value>
</property>
<property>
    <name>dfs.hosts.exclude</name>
    <value>/home/hadoop/hadoop2.6.5/etc/hadoop/exclude-slaves</value>
</property>
```

（3）下线 Hadoop 数据节点。执行节点刷新命令：

```
/home/hadoop/hadoop-2.5.6/bin/hadoop dfsadmin -refreshNodes
```

然后执行命令 hadoop dfsadmin -report 或者用浏览器打开链接 http://master1:50070，可以看到，该数据节点状态转为正在退役（Decommission In Progress），等退役进程完成数据迁移后，数据节点的状态会变成已退役（Decommissioned），然后数据节点进程会自动停止。此时节点 slave2 已经转移到下线节点（dead nodes）列表中。

（4）清理 slave2。将 slaves 与 exclude-slaves 文件中 slave2 这行数据删除，然后执行刷新 hadoop 节点命令，到此 slave2 即已经成功下线。

第 3 章

HBase 数据模型

HBase 的数据模型与传统数据库相比更加灵活，使用之前无须预先定义一个所谓的表模式（Schema），同一个表中不同行数据可以包含不同的列，而且 HBase 对列的数量并没有限制。当然如果一行包括太多的列，就会对性能产生负面影响。HBase 很适合存储不确定列、不确定大小的半结构化数据。

3.1 逻辑模型

HBase 是一个键值（key-value）型数据库。HBase 数据行可以类比成一个多重映射（map），通过多重的键（key）一层层递进可以定位一个值（value）。因为 HBase 数据行列值可以是空白的（这些空白列是不占用存储空间的），所以 HBase 存储的数据是稀疏的。

下面解释一下与 HBase 逻辑模型相关的名词。

（1）表（table）：类似于关系型数据库中的表，即数据行的集合。表名用字符串表示，一个表可以包含一个或者多个分区（region）。

（2）行键（row key）：用来标识表中唯一的一行数据，以字节数组形式存储，类似于关系型数据库中表的主键（不同的是从底层存储来说，行键其实并不能唯一标识一行数据，因为 HBase 数据行可以有多个版本。但是，一般在不指定版本或者数据时间戳的情况下，用行键可以获取到当前最新生效的这行数据，因此从用户视图来说，默认情况下行键能够标识唯一一行数据），同时行键也是 HBase 表中最直接最高效的索引，表中数据按行键的字典序排序。

（3）列族（column family）：HBase 是一个列式存储数据库，所谓列式就是根据列族存储，每个列族一个存储仓库（Store），每个 Store 有多个存储文件（StoreFile）用来存储实际数据。

（4）列限定符（column qualifier）：每个列族可以有任意个列限定符用来标识不同的列，这个列也类似于关系型数据库表的一列，与关系型数据库不同的是列无须在表创建时指定，可以在需要使用时动态加入。

（5）单元格（cell）：单元格由行键、列族、列限定符、时间戳、类型（`Put`、`Delete`等用来标识数据是有效还是删除状态）唯一决定，是 HBase 数据的存储单元，以字节码的形式存储。

（6）版本（version）：HBase 数据写入后是不会被修改的，数据的 `Put` 等操作在写入预写入日志（Write-Ahead-Log，WAL）（类似于 Oracle Redo Log）后，会先写入内存仓库（MemStore），同时在内存中按行键排序，等到合适的时候会将 MemStore 中的数据刷新到磁盘的 StoreFile 文件。因为数据已经排序，所以只需顺序写入磁盘，这样的顺序写入对磁盘来说效率很高。由于数据不会被修改，因此带来的问题就是数据会有多个版本，这些数据都会有一个时间戳用来标识数据的写入时间。

（7）分区（region）：当传统数据库表的数据量过大时，我们通常会考虑对表做分库分表。例如，淘宝的订单系统可以按买家 ID 与按卖家 ID 分别分库分表。同样 HBase 中分区也是一个类似的概念，分区是集群中高可用、动态扩展、负载均衡的最小单元，一个表可以分为任意个分区并且均衡分布在集群中的每台机器上，分区按行键分片，可以在创建表的时候预先分片，也可以在之后需要的时候调用 HBase shell 命令行或者 API 动态分片。

接下来以用户行为管理系统为例，假设现在需要用 HBase 来存储电商系统的用户行为数据，表 `s_behavior` 用来存储这些行为数据，两个列族 `pc` 和 `ph` 分别存储电脑端与手机端的行为数据，列 `v` 用来存储用户浏览记录，列 `o` 用来存储用户的下单记录，图 3-1 描述了这个表的 HBase 逻辑视图。

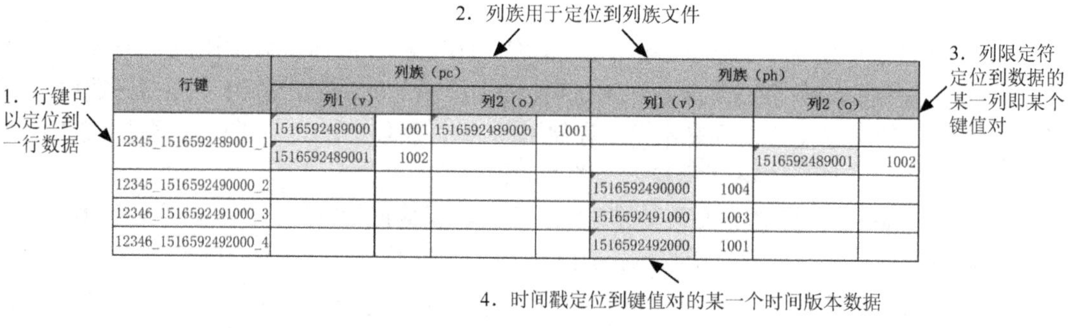

行键	列族（pc）		列族（ph）	
	列1（v）	列2（o）	列1（v）	列2（o）
12345_1516592489001_1	1516592489000 1001	1516592489000 1001		
	1516592489001 1002			1516592489001 1002
12345_1516592490000_2			1516592490000 1004	
12346_1516592491000_3			1516592491000 1003	
12346_1516592492000_4			1516592492000 1001	

1. 行键可以定位到一行数据
2. 列族用于定位到列族文件
3. 列限定符定位到数据的某一列即某个键值对
4. 时间戳定位到键值对的某一个时间版本数据

图 3-1　HBase 逻辑视图

HBase 按行键的字典序存储数据行，其数据存储层级可以用如下的一个 Java 中的 `Map` 结构类比：

`Map<RowKey,Map<Column Family,Map<Column Qualifier,Map<Timestamp,Value>>>>`

如下代码可以用来定位到某个键值对（或者说单元格）：

`map.get("12345_1516592489001_1").get("pc").get("v").get("1516592489001");`

代码清单 3-1 的 JSON 字符串同样近似地描述了这个多维的 `Map`。

代码清单 3-1　HBase 逻辑视图类比 JSON
```
{
    "12345_1516592489001_1": {
        "pc": {
            "pc:v": {
                "1516592489000": "1001",
                "1516592489001": "1002"
            },
            "pc:o": {
                "1516592489000": "1001"
            }
        },
        "ph": {
            "ph:v": {
                "1516592489001": "1002"
            }
        }
    },
    "12345_1516592490000_2": {
        "ph": {
            "ph:v": {
                "1516592490000": "1004"
            }
        }
    }
}
```

3.2　物理模型

HBase 是一个列式存储数据库，数据按列族聚簇存储在存储文件（StoreFile）中，空白的列单元格不会被存储，图 3-2 描述了 HBase 表的物理存储模型。

（1）HBase 中表按照行键的范围被划分为不同的分区（Region），各个分区由分区服务器负责管理并提供数据读写服务，HBase 主节点进程（HMaster）负责分区的分配以及在集群中的迁移。

（2）一个分区同时有且仅由一个分区服务器提供服务。当分区增长到配置的大小后，如果开启了自动拆分（也可以手动拆分或者建表时预先拆分），则分区服务器会负责将这个分区拆分成两个。每个分区都有一个唯一的分区名，格式是"<表名,startRowKey,创建时间>"。一个分区下的每个列族都会有一个存储仓库（Store），因此一个表有几个列族，那么每个分区就会有几个存储仓库。

（3）每个 Store（存储仓库）有且仅有一个 MemStore（内存仓库），但是可以有多个存储文件。当分区服务器处理写入请求时，数据的变更操作在写入 WAL 后，会先写入 MemStore，同时在内存中按行键排序。当 MemStore 到达配置的大小或者集群中所有 MemStore 使用的总内存达到配置的阈值百分比时，MemStore 会刷新为一个 StoreFile（存储文件）到磁盘，存储文件只会顺序写入，不支持修改。

（4）数据块（block）是 HBase 中数据读取的最小单元，StoreFile 由数据块组成，可以在建表时按列族指定表数据的数据块大小。如果开启了 HBase 的数据压缩功能，数据在写入 StoreFile 之前会按数据块进行压缩，读取时同样对数据块解压后再放入缓存。理想情况下，每次读取数据的大小都是指定的数据块大小的倍数，这样可以避免一些无效的 IO，效率最高。

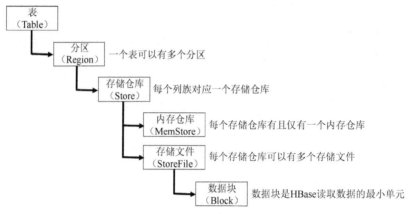

图 3-2　HBase 物理视图

图 3-3 描述了上面提到的表和分区等各模块分别由 HBase 的哪些进程负责管理。

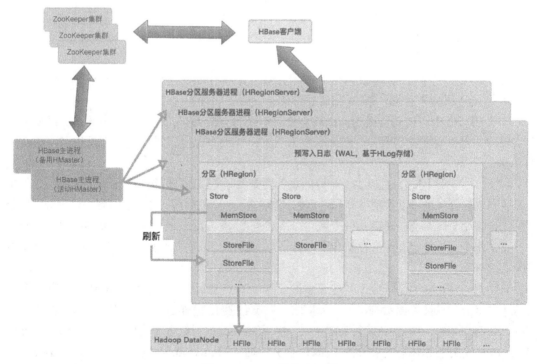

图 3-3　HBase 模块交互图

- HMaster：负责监控集群中所有的分区服务器进程（HRegionServer），负责所有元数据的更新（如分区由哪个分区服务器提供服务）。HMaster 同时负责分区在分区服务器中的负载均衡，在一个分布式集群中，HMaster 进程通常与 Hadoop 的 NameNode 运行在同一个节点，每个集群会部署至少两个 HMaster 节点，一个作为活跃节点提供服务，另一个作为备用节点提供快速的灾备切换，保证集群的高可用。
- HRegionServer：管理其负责的分区，处理分区的读写请求、分区增大的拆分（split）以及分区的压缩（compact）。HBase 客户端根据元数据（客户端从 HMaster 获取到元数据后会缓存在本地，当分区操作抛出特定异常后会从 HMaster 刷新缓存）定位到操作数据对应的分区所在分区服务器之后，HBase 客户端对数据的读写就直接与分区服务器交互，因此对分区的读写不会对 HMaster 造成压力。HRegionServer 进程通常与 Hadoop 的 DataNode 运行在同一个节点，这样对数据的读取可以尽量做到本地读取，减少网络请求。
- WAL：默认情况下一个分区服务器仅有一个 WAL。HBase 客户端数据请求操作会先写入 WAL 文件再写入内存仓库 MemStore，这样当分区服务器宕机重启的时候，可以用 WAL 来恢复分区服务器的状态（如 MemStore 中更新的数据没有刷新到 StoreFile 持久化，则分区服务器启动时需要通过 WAL 重做（replay）数据更新来恢复）。如果数据操作写入 WAL 失败，则这个更新数据的请求也会失败。由于一个分区服务器只有一个 WAL，对 WAL 的写入必须是顺序写入，这里可能导致性能瓶颈，HBase1.0 引入了 mutiWAL 的支持，mutiWAL 允许分区服务器并发地写入多个 WAL，这个并发来自于对 WAL 更新按分区分组，也就是说不同分区可以支持并发写多个 WAL，因此 mutiWAL 无法提升对单个分区的高并发更新的性能。
- Store：每个分区的每个列族对应一个存储仓库，一个存储仓库包含一个 MemStore 和多个存储文件。当 MemStore 的大小达到了配置的阈值后，MemStore 会刷新为一个存储文件，存储文件顺序写入，不支持修改，以 HFile 的形式存储在 Hadoop 的 DataNode 中。
- MemStore：MemStore 位于分区服务器的堆内存，数据在写入 MemStore 的时候即会按行键排序，这样刷新到存储文件的时候可以直接顺序写入，提高写性能。同时，MemStore 作为一个内存级缓存，能够提供对新写入数据的快速访问（新插入数据总是比老数据使用频繁）。

第 4 章

HBase shell

HBase 提供了一个非常方便的命令行交互工具 HBase shell。通过 HBase shell，HBase 可以与 MySQL 命令行一样创建表、索引，也可以增删查数据，同时集群的管理、状态查看等也可以通过 HBase shell 实现。

下面以用户行为日志系统数据存储表 s_behavior 为例介绍相关命令。假设该表用来存储用户手机端和计算机端的行为数据，列族 pc 用来存储计算机端的行为数据，列族 ph 用来存储手机端的行为数据，每个列族都有两列，列 1（v）代表用户的商品浏览记录，列 2（o）代表用户的商品下单记录，行键格式为"[用户 ID]_[时间戳]_[序列号]"，表数据如图 4-1 所示。

行键	列族（pc）		列族（ph）	
	列1（v）	列2（o）	列1（v）	列2（o）
12345_1516592489001_1	1001			1001
12345_1516592490000_2			1004	
12346_1516592491000_3			1003	
12346_1516592492000_4			1001	

图 4-1 用户行为日志表

4.1 数据定义语言

数据定义语言（Data Definition Language，DDL），包括数据库表的创建、修改等语句。

4.1.1 创建表

创建表的语句如下：

```
create 's_behavior', {NAME => 'pc'} , {NAME => 'ph'}
```

该语句创建了一个 s_behavior 表,用来存储用户的行为数据,这个表有两个列族,列族 pc 用来存储用户 PC 端的用户行为数据,列族 ph 用来存储用户手机端的用户行为数据。

4.1.2 查看所有表

查看所有表的语句如下:

```
list
```

该语句会列出 HBase 数据库中所有已经创建的表。

4.1.3 查看建表

查看建表的语句如下:

```
describe 's_behavior'
```

该语句的输出结果如图 4-2 所示。

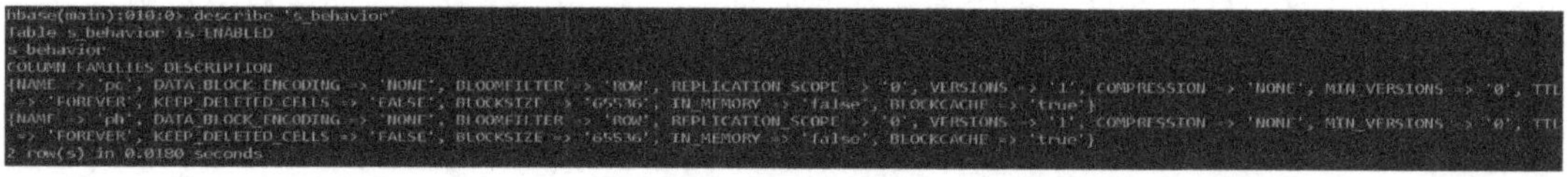

图 4-2 查看建表语句

可以看到虽然在建表的时候没有指定任何属性,但是 HBase 默认会给表设置一些属性,下面逐个解释这些属性。

- DATA_BLOCK_ENCODING:数据块编码。用类似于压缩算法的编码形式来节省存储空间,主要是针对行键,用时间换空间,默认情况下不启用数据块编码,第 9 章会有详细描述。
- BLOOMFILTER:布隆过滤器。数据查询 Scan 操作的时候用来排除待扫描的 StoreFile 文件,第 9 章会有详细描述。
- REPLICATION_SCOPE:集群间数据复制开关。当集群间数据复制配置好后,REPLICATION_SCOPE=1 的表会开启复制。默认为 0,表示不开启复制。
- VERSIONS:HBase 对表的数据行可以保留多个数据版本,以时间戳来区分。VERSIONS 表示对该表 HBase 应该保留多少个数据版本。
- COMPRESSION:压缩方式。HBase 提供了多种压缩方式用来在数据存储到磁盘之前压缩以减少存储空间，例如 SNAPPY、LZO、GZIP 等。第 9 章也会详细介绍并对比各种压缩方式。
- TTL：指 Time To Live，数据的有效时长。超过有效时长的数据在主压缩（或者主合并，major compact）的时候会被删除。
- KEEP_DELETED_CELLS：保留了删除的数据。意味着可以通过 Get 或者 Scan 请求获取已经被删除了的数据（如果数据删除后经过了一次主压缩，那么这些删除的数据也会被清理），只要这行数据的时间戳落在查询的时间范围即可。注意，如果需要开启集群间复制，则这个属性必须为 true，否则可能导致数据复制失败。
- BLOCKSIZE：HBase 读取数据的最小单元。设置过大会导致读取很多不需要的数据，过小则会产生更多的索引文件，默认大小为 64 KB。

4.1.4 修改表

修改表的模式（schema）之前需要将表先下线，然后执行修改的命令，再上线，代码清单 4-1 将 s_behavior 表修改为开启集群间复制。

代码清单 4-1 修改表
```
disable 's_behavior'
alter 's_behavior', {NAME=>"cf", REPLICATION_SCOPE=>"1", KEEP_DELETED_CELLS => 'TRUE'}
enable 's_behavior'
```

4.2 数据操纵语言

数据操纵语言（Data ManipulationLanguage，DML），包括数据的修改、查询、删除等语句。

4.2.1 Put

Put 命令用来插入一行数据到 HBase 表，命令格式如下：

```
put <table>,<rowkey>,<列族:列标识符>,<值>
```

如代码清单 4-2 所示，第一条命令插入一条用户 PC 端行为数据，第二条命令插入一条用户手机端行为数据。注意，两条命令行键相同，只是指定了不同的列族，因此实际上两条数据在 HBase 中逻辑上是同一条数据。之后用该行键可以唯一地定位到这两个列族的数据，接下来的 Get 和 Scan 命令可以验证这一点。

代码清单 4-2　插入数据

```
hbase(main):004:0> put 's_behavior','12345_1516592489001_1','pc:v','1001'
0 row(s) in 0.2020 seconds

hbase(main):005:0> put 's_behavior','12345_1516592489001_1','ph:o','1001'
0 row(s) in 0.0050 seconds
```

4.2.2　Get

Get 命令用来根据行键获取 HBase 表的一条记录，命令格式如下：

```
get <table>,<rowkey>
```

如代码清单 4-3 所示，使用 Get 命令通过行键获取表 s_behavior 的一行数据，包括这行数据的所有列族。

代码清单 4-3　获取一条数据

```
hbase(main):006:0> get 's_behavior','12345_1516592489001_1'
COLUMN                    CELL
 pc:v                     timestamp=1521423996693, value=1001
 ph:o                     timestamp=1521423996739, value=1001
2 row(s) in 0.0330 seconds
```

由前面的建表 DDL 可知 HBase 支持数据行的多版本，Get 命令可以通过指定时间戳来获取一行数据某个时刻的镜像，如代码清单 4-4 所示，根据时间戳获取一行数据在该时刻的数据。

代码清单 4-4　获取一条数据某时刻的镜像

```
hbase(main):007:0> get 's_behavior','12345_1516592489001_1' , {TIMESTAMP => '1521423996693'}
COLUMN                    CELL
 pc:v                     timestamp=1521423996693, value=1001
1 row(s) in 0.0120 seconds
```

Get 命令也支持获取数据的多个版本，但是需要在建表语句里面指定 VERSIONS 属性，如代码清单 4-5 所示，先修改表支持多 VERSIONS，然后用 Get 获取多个版本的数据。

代码清单 4-5　获取多个版本的数据

```
hbase(main):008:0> alter 's_behavior',NAME=>'pc',VERSIONS =>3
Updating all regions with the new schema...
1/1 regions updated.
Done.
0 row(s) in 1.9370 seconds

hbase(main):009:0> put 's_behavior','12345_1516592489001_1','pc:v','1002'
0 row(s) in 0.0100 seconds
```

```
hbase(main):010:0> get 's_behavior','12345_1516592489001_1',{COLUMN => 'pc:v',VERSIONS=>2}
COLUMN                                      CELL
 pc:v                                       timestamp=1521424961757, value=1002
 pc:v                                       timestamp=1521423996693, value=1001
2 row(s) in 0.0240 seconds
```

4.2.3 Scan

Scan 命令用来扫描表的数据。Scan 是 HBase 数据查询命令中最复杂的命令，需要特别注意查询的数据量，以免由于扫描数据过大导致 HBase 集群出现响应延迟，Scan 命令格式如下：

```
scan <table>
```

如代码清单 4-6 所示，使用 Scan 命令来查询表 s_behavior 的所有记录。

代码清单 4-6　扫描表数据

```
hbase(main):012:0* scan 's_behavior'
ROW                            COLUMN+CELL
 12345_1516592489001_1         column=pc:v, timestamp=1521424961757, value=1002
 12345_1516592489001_1         column=ph:o, timestamp=1521423996739, value=1001
1 row(s) in 0.0120 seconds
```

Scan 命令可以通过指定时间区间获取某个时刻的数据镜像，也可以指定 VERSIONS 以获取多个版本的数据。Scan 同样可以使用各种过滤器（Filter，类似 MySQL 的 where 条件语句）。Scan 操作不论是在 HBase shell 还是在客户端 API，都是最复杂的一个操作，如果使用不当，就可能造成 HBase 集群的响应延迟。

（1）获取时间区间内数据。如代码清单 4-7 所示，使用时间区间获取了"1521423996739<= 时间戳<1521424961757"的数据。

代码清单 4-7　获取时间区间内数据

```
hbase(main):013:0>  scan 's_behavior', {TIMERANGE => [1521423996739,1521424961757]}
ROW                            COLUMN+CELL
 12345_1516592489001_1         column=ph:o, timestamp=1521423996739, value=1001
1 row(s) in 0.0190 seconds
```

（2）获取多版本数据。如代码清单 4-8 所示，使用 VERSIONS 参数获取表的两个版本数据。

代码清单 4-8　获取两个版本数据

```
hbase(main):003:0* scan 's_behavior',{VERSIONS=>2}
ROW                            COLUMN+CELL
 12345_1516592489001_1         column=pc:v, timestamp=1521424961757, value=1002
```

```
 12345_1516592489001_1          column=pc:v, timestamp=1521423996693, value=1001
 12345_1516592489001_1          column=ph:o, timestamp=1521423996739, value=1001
1 row(s) in 0.1200 seconds
```

（3）获取用户（ID=12345）前 5 行数据。因为 s_behavior 表的行键都是以用户 ID 开头，所以这里可以使用一个前缀过滤器，注意需要在用户 ID 后面带上下划线，否则会匹配到行键诸如 123450_xxx 等类似的数据，如代码清单 4-9 所示。

代码清单 4-9　获取用户前 5 行数据

```
hbase(main):013:0* scan 's_behavior', {FILTER => "PrefixFilter('12345_')" ,COLUMNS => ['pc'],LIMIT=>5}
ROW                             COLUMN+CELL
 12345_1516592489001_1          column=pc:v, timestamp=1521442781396, value=1002
 12345_1516592489001_2          column=pc:v, timestamp=1521442781447, value=1003
 12345_1516592489001_3          column=pc:v, timestamp=1521442781471, value=1004
 12345_1516592489001_4          column=pc:v, timestamp=1521442781492, value=1005
 12345_1516592489001_5          column=pc:v, timestamp=1521442781513, value=1006
5 row(s) in 0.0240 seconds
```

（4）获取用户（ID=12345）某个时间区间的 PC 端行为数据。Scan 可以指定扫描开始和结束行键。因为 s_behavior 表行键的第二个因子是时间戳，所以只需将需要扫描的时间范围转化为开始和结束行键即可，如代码清单 4-10 所示，使用 STARTROW、STOPROW 参数可以指定扫描一个先开后闭的区间，即"STARTROW<=行键<STOPROW"。

代码清单 4-10　获取用户某个时间区间内产生的 PC 端行为数据

```
hbase(main):014:0> scan 's_behavior', {STARTROW =>
'12345_1516592489001' ,STOPROW=>'12345_1516592489002' ,COLUMNS => ['pc']}
ROW                             COLUMN+CELL
 12345_1516592489001_1          column=pc:v, timestamp=1521442781396, value=1002
 12345_1516592489001_2          column=pc:v, timestamp=1521442781447, value=1003
 12345_1516592489001_3          column=pc:v, timestamp=1521442781471, value=1004
 12345_1516592489001_4          column=pc:v, timestamp=1521442781492, value=1005
 12345_1516592489001_5          column=pc:v, timestamp=1521442781513, value=1006
 12345_1516592489001_6          column=pc:v, timestamp=1521442781533, value=1007
 12345_1516592489001_7          column=pc:v, timestamp=1521442781553, value=1008
 12345_1516592489001_8          column=pc:v, timestamp=1521442782603, value=1009
8 row(s) in 0.0320 seconds
```

（5）获取用户对商品（ID=1001）的行为数据。ValueFilter 可以用来限定某个列的值等于指定的值，也可以用来限定某个列的值包含某个值。如代码清单 4-11 所示，第一条命令查询表 s_behavior 列值等于 1001 的数据，第二条命令查询表 s_behavior 列值包含 1002 的数据。

代码清单 4-11　获取用户对商品 1001 的行为数据

```
hbase(main):015:0> scan 's_behavior', FILTER=>"ValueFilter(=,'binary:1001')"
ROW                             COLUMN+CELL
```

```
 12345_1516592489001_1          column=pc:v, timestamp=1521423996693, value=1001
 12345_1516592489001_1          column=ph:o, timestamp=1521423996739, value=1001
1 row(s) in 0.0430 seconds

hbase(main):016:0> scan 's_behavior', FILTER=>"ValueFilter(=,'substring:1002')"
ROW                             COLUMN+CELL
 12345_1516592489001_1          column=pc:v, timestamp=1521442781396, value=1002
1 row(s) in 0.0200 seconds
```

细心的读者会注意到 ValueFilter 会匹配列族下面所有的列,假如只需要查询用户对商品 1001 的下单行为数据该如何查询呢?答案是 SingleColumnValueFilter 可以指定搜索的列限定符,如代码清单 4-12 所示。

代码清单 4-12　获取用户对商品 1001 的下单行为数据

```
hbase(main):001:0> import org.apache.hadoop.hbase.filter.SingleColumnValueFilter;
hbase(main):002:0* import org.apache.hadoop.hbase.filter.BinaryComparator;
hbase(main):003:0* import org.apache.hadoop.hbase.filter.CompareFilter;
hbase(main):004:0* import org.apache.hadoop.hbase.util.Bytes;
hbase(main):005:0> scan 's_behavior',{FILTER =>
SingleColumnValueFilter.new(Bytes.toBytes('ph'), Bytes.toBytes('o'),
CompareFilter::CompareOp.valueOf('EQUAL'),
BinaryComparator.new(Bytes.toBytes('1001'))),COLUMNS => ['ph:o']}
ROW                             COLUMN+CELL
 12345_1516592489001_1   column=ph:o, timestamp=1521423996739, value=1001
1 row(s) in 0.0280 seconds
```

(6) 获取用户 (ID=12345) 行为数据的行键。过滤器通过括号、AND 和 OR 的条件组合支持多个过滤条件,使用 KeyOnlyFilter 可以使得 HBase 服务端只返回数据的行键以减少传输的数据量,如代码清单 4-13 所示。

代码清单 4-13　获取用户行为数据行键

```
hbase(main):012:0> scan 's_behavior', {FILTER => "PrefixFilter('12345') AND
KeyOnlyFilter()" , LIMIT=>3}
ROW                             COLUMN+CELL
 12345_1516592489001_1          column=pc:v, timestamp=1521442781396, value=
 12345_1516592489001_1          column=ph:o, timestamp=1521423996739, value=
 12345_1516592489001_2          column=pc:v, timestamp=1521442781447, value=
 12345_1516592489001_3          column=pc:v, timestamp=1521442781471, value=
3 row(s) in 0.0470 seconds
```

4.2.4　删除数据

HBase 提供了 Delete、DeleteAll 和 truncate 命令分别用来删除列、行和表的数据。
(1) 删除某列数据。格式如下:

```
delete '<table>', '<rowkey>', '<列族:列标识符>', '[<time stamp>]'
```

代码清单4-14所示命令删除行键为 12345_1510720956000_1 的数据行中列族为 ph、列限定符为 o 的数据。

代码清单 4-14　删除某列数据

```
hbase(main):014:0> delete 's_behavior','12345_1516592489001_1' ,'ph:o'
0 row(s) in 0.0470 seconds
```

（2）删除整行数据。格式如下：

```
deleteall '<table>', '<rowkey>
```

代码清单4-15所示命令删除了行键为 12345_1510720956000_1 的数据行。

代码清单 4-15　删除整行数据

```
hbase(main):015:0> deleteall 's_behavior','12345_1516592489001_1'
0 row(s) in 0.0090 seconds
```

（3）删除整表数据。格式如下：

```
truncate '<table>'
```

代码清单4-16所示命令清除了用户行为表 's_behavior' 所有的数据，注意清除表数据之前会将表先禁用。

代码清单 4-16　删除整表数据

```
hbase(main):016:0> truncate 's_behavior'
Truncating 's_behavior' table (it may take a while):
 - Disabling table...
 - Truncating table...
0 row(s) in 3.4640 seconds
```

4.3　其他常用 shell

HBase shell 还提供了非常丰富的命令用来管理和查看 HBase 状态，下面介绍一些常用的集群管理查看命令。

4.3.1　复制状态查看

当 HBase 开启了集群间复制时，使用 status 命令可以查看复制的状态，包括复制延迟、待复制的日志文件队列大小等，如代码清单4-17所示。

代码清单 4-17　复制状态查看

```
hbase(main):017:0> status 'replication'
version 1.2.6
3 live servers
    master1:
        SOURCE: PeerID=1, AgeOfLastShippedOp=4773835960, SizeOfLogQueue=2677,
        TimeStampsOfLastShippedOp=Thu Jan 01 08:00:00 CST 1970,
        Replication Lag=1521448279318
        SINK  : AgeOfLastAppliedOp=0,
        TimeStampsOfLastAppliedOp=Mon Dec 25 17:23:16 CST 2017
    slave1:
        SOURCE: PeerID=1, AgeOfLastShippedOp=5986238820, SizeOfLogQueue=1665,
        TimeStampsOfLastShippedOp=Thu Jan 01 08:00:00 CST 1970,
        Replication Lag=1521448277282
        SINK  : AgeOfLastAppliedOp=0,
        TimeStampsOfLastAppliedOp=Mon Dec 25 11:10:14 CST 2017
    master2:
        SOURCE: PeerID=1, AgeOfLastShippedOp=24121908, SizeOfLogQueue=7,
        TimeStampsOfLastShippedOp=Mon Mar 19 09:46:36 CST 2018,
        Replication Lag=24281038
        SINK  : AgeOfLastAppliedOp=0,
        TimeStampsOfLastAppliedOp=Tue Mar 13 17:02:06 CST 2018
```

4.3.2　分区拆分

实时在线集群一般会禁用自动拆分以免影响性能，因此当 StoreFile 大小到达某个值后需要手动或者使用自动化程序将分区做拆分。例如，代码清单 4-18 将用户行为表 s_behavior 拆分为两个分区，由于行键以用户 ID 开始，而用户 ID 第一个字符取值范围为 0～9，中值为 5，因而可以使用 5 作为拆分后两个分区的分割字符。

代码清单 4-18　分区拆分

```
hbase(main):018:0>  split 's_behavior','5'
0 row(s) in 0.1060 seconds
```

4.3.3　分区主压缩

实时在线集群同样也会禁止主压缩（major compact）或者主合并而等到请求的非高峰期来定时做主压缩，代码清单 4-19 手动触发主压缩来压缩 s_behavior 表的分区，分区名称可以在 HBase Web UI 找到。

代码清单 4-19　分区压缩
```
hbase(main):040:0> major_compact
's_behavior,,1511878479015.e933a5867bd5253211a4ef90e549192f.'
0 row(s) in 0.0200 seconds
```

4.3.4　负载均衡开关

如果需要对集群进行滚动升级或者想要关闭自动负载均衡而采用手动负载均衡模式，则可以使用代码清单 4-20 所示命令来关闭或者开启自动负载均衡。

代码清单 4-20　开启关闭负载均衡
```
hbase(main):019:0> balance_switch true
true
0 row(s) in 0.0220 seconds

hbase(main):020:0> balance_switch false
true
0 row(s) in 0.0090 seconds
```

4.3.5　分区手动迁移

如果有几个比较大或者负载高的分区被分配到同一个分区服务器，那么这台分区服务器可能会是整个 HBase 集群的瓶颈，这时候可以手动将这些分区迁移到负载低的分区服务器。

分区迁移命令格式如下：

move' <EncodedRegionName>','<destSeverName>'

注意，EncodedRegionName 是 RegionName 的后缀，destServerName 则是在 HBase Web UI 上面的分区服务器全名，如代码清单 4-21 所示。

代码清单 4-21　分区迁移
```
hbase(main):002:0> move 'e933a5867bd5253211a4ef90e549192f', 'master2,16020,15130
49558323'
0 row(s) in 0.0790 seconds
```

第 5 章 模式设计

虽然 HBase 是一个"无模式"的数据库，但是当架构一个系统时，仍然得考虑为了满足系统功能需求应该设计几个表、每个表存储什么类型的数据以及如何优化对数据的查询。很多时候应该在设计的时候就考虑优化，而不是等系统开发完成或者数据量已经庞大到无法动弹的时候再来考虑优化。传统的关系型数据库在设计表的同时需要考虑如何给表添加索引，类似地，HBase 在设计表的同时应该考虑如何设计行键以提高查询效率。行键在 HBase 中充当表的一级索引角色，并且 HBase 本身没有提供二级索引的机制，因此对行键的设计优化对实时查询尤为重要。

MySQL 等关系型数据库在写数据的同时需要随机写磁盘来构建索引，而 HBase 系统架构则通过写内存、顺序写磁盘来提升写性能（磁盘顺序写性能比随机写性能高很多），但这是以牺牲读性能为代价的。因为内存中的数据最终会刷新为 StoreFile，所以最后会有很多个 StoreFile 来存储数据，读取数据时也就需要读取多个 StoreFile 来查找到所需要的数据。同样也有许多方法可以用来提升读性能，例如通过写多份数据来优化来自不同维度的数据查询，如电商订单的卖家维度与买家维度数据，也可以通过牺牲一些业务可用性来提升性能，如客户端缓存，禁写 WAL 可以使得写速度更快，但是可能会导致一些数据的丢失。

下面以用户行为日志系统为例来设计其 HBase 存储方案，假设我们的系统仍然聚焦在电商业务，数据类型、数据特征与数据分析需求如下所示。

数据类型：
- 用户商品浏览记录。
- 用户商品下单记录。

统计需求：
- 需要查询同一个用户一段时间内浏览过的商品，用来做用户商品推荐以提高转化率。
- 需要统计某个商品在某天的转化率,用来分析各渠道或者营销方式的优劣。

数据特征:
- 数据一旦写入就不会被修改。
- 越旧的数据越少使用,越新的数据越多使用。
- 数据通常需要被大批量读取用作分析。

5.1 行键设计

HBase 数据按照行键字典序自然排序,这对扫描(Scan)操作是一个优化。行键也是 HBase 最有效的索引,与 MySQL 之类的传统关系型数据库支持多字段的索引不同的是,HBase 不支持二级索引,因此 MySQL 通过多重、多列索引支持的复杂数据库查询操作对 HBase 来说可能是个灾难。当然 HBase 也支持对列的条件过滤(参见 6.4 节),但是因为需要读取存储文件做字符或者字节对比,所以效率很低,对性能影响很大。

下面列出了一些 HBase 行键设计的原则。
- 唯一原则:行键对应关系型数据库的唯一键,系统设计之初必须考虑有足够的唯一行键去支持业务的数据量。
- 长度原则:长度适中,一般从几十到一百字节,建议使用定长,方便从行键提取所需数据,而无须查询出数据内容以节省网络开销。
- 散列原则:避免递增,否则读写负载都会集中在某个热点分区,降低性能,甚至引起分区服务器过载而宕机。

下面从统计需求来看用户行为日志系统表 s_behavior 的行键设计。
- 查询同一个用户一段时间内浏览过的商品:该需求有两个数据查询维度,第一个是用户,第二个是时间。为了提高查询性能,需要把同一个用户的数据聚簇地放在一起,因此可以把用户 ID 作为行键的开始。类似于 MySQL 的组合索引,用户 ID 作为组合索引的引导列,时间则可以作为组合索引的第二列,这样得到的行键格式为"[用户 ID]_[时间戳]"。在某些极端并发情况下,例如,用户浏览器同时打开多个商品,那么这些浏览记录时间戳可能相同。考虑到唯一性原则,可以在行键最后添加一个序列号来实现,因此最后得到的行键格式为"[用户 ID]_[时间戳]_[序列号]"。用户浏览的商品 ID、订单 ID 等可以作为列值数据存储。
- 统计某个商品某天的转化率:转化率的定义是商品下单数除以商品浏览数,该需求有 3 个数据查询维度,第一个是商品,第二个是数据类型,第三个是时间。可以把商品 ID 作为行键的开始,接下来用一个数字表示用户的数据类型(如 0:商品浏览记录;1:商品下单记录),再加上时间戳。同一个商品同一时间会有很多用户浏览和下单，根据唯一性原则，行健最后也需要添加一个序列号，因此最后得到的行键格式为"[商品 ID]_[数据类型]_[时间戳]_[序列号]"，用户 ID 等可作为列值数据存储。

当然一个完整的用户日志系统还要考虑很多其他需求，但是到目前为止的两个需求已经足够麻烦了，因为两个需求得到的行键完全没有共同点。下面分析如何在保持性能基本稳定的同时满足以上两个需求。

- 使用时间戳过滤：注意，两个需求的行键都包括了时间戳，那么是否可以把时间戳顺序提前用作行键的开始呢？假如把时间戳用作行键，那么无论统计哪个需求，都需要把整个时间段内所有的数据都扫描出来，然后在服务端或者客户端过滤，这样会导致大量非必要的数据读取。例如，查询某个用户在 2018 年 1 月 1 日这一天所有的商品浏览数据，则需要将整个用户行为系统在这一天产生的所有数据全部扫描一次再过滤，性能无疑非常低下，无法满足需求。
- 使用二级索引：HBase 本身不支持二级索引，唯一的索引就是行键。由于 HBase 对复杂查询过滤条件支持的局限性，因此开源社区对 HBase 二级索引提出了多个解决方案，如 IHBase、华为的 HIndex 等。这些解决方案可能实现方式不一样（如华为 HIndex 使用协处理器实现），各有优劣，但是最终目标都是实现 HBase 对复杂查询条件的支持，同时在性能、数据一致性、代码侵入性方面能够达到平衡。

回到用户行为管理系统，为了满足两个查询需求，可以考虑按用户维度构建行键存储数据，按商品维度建立二级索引满足商品维度统计需求，如表 5-1 所示。

表 5-1 用户行为日志管理系统二级索引设计

行键	列族:列浏览记录 （cf:v）	列族:列下单记录 （cf:o）
12345_1510720956000_1	1001	1001
12345_1510721056000_2	1002	
12346_1510721086000_3	1001	
12346_1510721096000_4		1001
...	...	...
12345_idx1_1001_0_1510720956000_12345_1510720956000_1		
12345_idx1_1001_0_1510721086000_12346_1510721086000_3		
12345_idx1_1001_1_1510720956000_12345_1510720956000_1		
12345_idx1_1001_1_1510721096000_12346_1510721096000_4		
12345_idx1_1002_0_1510721086000_12345_1510721056000_2		

表 5-1 前 4 行为数据部分，后 4 行为索引部分，下面解释每行数据的含义。

- 第一行：用户 12345 在时间戳 1510720956000 浏览了商品 1001，同时下单了商品 1001。
- 第二行：用户 12345 在时间戳 1510721056000 浏览了商品 1002。
- 第三行：用户 12346 在时间戳 1510721086000 浏览了商品 1001。
- 第四行：用户 12346 在时间戳 1510721096000 下单了商品 1001。
- 从第六行开始就是对前四行数据的索引，索引数据只有行键，列值数据均为空，索引格式为"[分区开始行键]_[索引名称]_[商品 ID]_[数据类型]_[时间戳]_[数据行键]"，使用分区开始行键作为行键的前缀是为了让索引和数据在同一个分区，这样读取索引和数据的请求就会落在同一个分区服务器，否则可能需要在分区服务器 A 读取索引后，再跨机器去分区服务器 B 读取数据。

下面看看如何使用二级索引满足系统的两个统计需求。

- 查询同一个用户一段时间内浏览过的商品：数据的行键以用户 ID 开始，因此天然支持对该用户数据的统计查询，扫描请求只需指定该用户的最小行键和最大行键。例如，用户查询用户 12345 从 2017 年 12 月 1 日到 2018 年 1 月 1 日之间的商品浏览记录，只需指定查询的行键区间[12345_1512086400000, 12345_1514764800000)。
- 统计某个商品某天的转化率：这里二级索引就可以发挥作用了，查询的主要维度是商品，因此可以用到以商品 ID 开始的索引 idx1。假设需要统计商品 1001 在 2018 年 1 月 1 日这天的转化率，需要扫描的行键区间为[12345_idx1_1001_0_1514764800000, 12345_idx1_1001_1_1514851200000)，并且只需扫描所有的索引行键。拿到所有的索引行键后循环一遍，根据数据类型判断是浏览记录还是下单记录，对浏览记录和下单记录，分别求和即可得到转化率。注意，这里的场景比较特殊，通常情况下还会需要根据索引行键得到数据行键，再根据数据行键去查询数据，因此索引行键的最后是数据行键。

5.2 规避热点区间

生产环境中用户 ID 的生成规则通常是一个递增的数字（用户管理系统一般都基于关系型数据库，用户 ID 一般使用 MySQL 等数据库提供的自增主键）。以用户行为日志表为例，假设将该表分为 10 个分区，分区行键范围分别是[0, 1), [1, 2), [2, 3), … [8, 9), [9,)，用户行为日志管理系统用户 ID 使用的递增数字计数器已经到了 12346，接下来的新注册用户 ID 会是 12347, 12348, …, 12366, …, 12398，而新注册的用户一般会比较活跃，会产生比较多的行为日志数据。这些数据的行键以用户 ID 开始，即行键第一个字符为 1，因此这些数据会落在分区[1,2)，显然接下来分区[1,2)会承受很大的读写压力，继而引起负责该分区的分区服务器负载升高，使得该分区服务器上的其他分区响应延长，最终可能会影响集群整体性能。同样一些使用时间戳作为行键的设计也会引起同样的问题，因此行键的设计需要规避这些问题，使得数据均匀、平衡地分布在集群的每台分区服务器。

下面是一些常见的避免热点区间的方法。

（1）加盐。在行键前面添加随机数字或者字母，使得数据随机分配到不同的分区，这种方式弊端显而易见。如果需要使用 GET 请求再次获取某行数据，则需要在插入时保存一个原始业务行键与添加的随机数的映射关系，或者使用某种散列函数计算原始业务行键的散列值，然后将该散列值作为最终行键的前缀。使用这种行键设计的应用通常只是用来做一些分析统计，因此一般实时在线系统不建议使用该行键设计方式。

（2）反转补齐。将用户 ID 反转（如 12345 反转为 54321）可以将变化最多的部分放到行键前面，这样数据的写入也能够顺序地流入各个分区而使得集群负载比较均衡。反转补齐是避免热点区间常用的方法。因为用户 ID 一般都使用关系型数据库的自增主键，长度最长一般为 20 个数字，所以为了使得行键保持定长以方便排序以及可以从行键反推出用户 ID，通常会将用户 ID 反转后在末尾加 0 补齐 20 个数字。类似地，如果使用时间戳作为行键的一部分，则可以使用 "Long.MAX_VALUE-时间戳"，这样最新时间戳的数据行键值较小，数据行能够排在数据存储文件前列，代码清单 5-1 实现了这两种行键设计的处理方式。

代码清单 5-1　行键反转补齐

```
1.    package com.mt.hbase.chpt5;
2.
3.    public class RowKeyUtil {
4.
5.        /**
6.         * 补齐20位，再反转
7.         *
8.         * @param userId
9.         * @return
10.        */
11.       public String formatUserId(long userId) {
12.           String str = String.format("%0" + 20 + "d", userId);
13.           StringBuilder sb = new StringBuilder(str);
14.           return sb.reverse().toString();
15.       }
16.
17.       /**
18.        * Long.MAX_VALUE-lastupdate 得到的值再补齐20位
19.        *
20.        * @param lastupdate, eg: "1479024369000"
21.        * @return
22.        */
23.       public String formatLastUpdate(long lastupdate) {
24.           if(lastupdate < 0){
25.               lastupdate = 0;
26.           }
```

```
27.            long diff = Long.MAX_VALUE - lastupdate;
28.            return String.format("%0" + 20 + "d", diff);
29.        }
30.
31.        public static void main(String [] args){
32.
33.            RowKeyUtil rowKeyUtil = new RowKeyUtil();
34.            // 下行输出结果为 54321000000000000000
35.            System.out.println(rowKeyUtil.formatUserId(12345L));
36.
37.            long time = System.currentTimeMillis();
38.            // 运行时下行输出结果为 09223370520503124434
39.            System.out.println(rowKeyUtil.formatLastUpdate(time));
40.
41.        }
42.
43.
44.    }
```

5.3 高表与宽表

用户行为日志表 s_behavior 的每行代表用户的一条浏览或者下单记录，每行最多包含两列，分别代表浏览或者下单记录。每当用户有新的浏览记录或者下单记录时，就在表中新增一行记录，最后这个表的数据会看起来"高高瘦瘦"的，称之为高表。与高表相反的还有另外一种设计，假设仍然以用户 ID 为查询维度，同样需要把同一个用户的数据聚簇地存储在一起。如果以用户 ID 为行键，则用户的每条浏览记录或者下单记录为数据行的一列，每当用户有新的浏览记录或者下单记录时，就更新表中以用户 ID 为行键的这行数据，对这行数据增加一列，最后这个表的数据会看起来"矮矮胖胖"的，称之为宽表。下面定义了高表与宽表。

- 高表（tall-narrow table）：数据每行包含的列比较少而行比较多的表。
- 宽表（flat-wide table）：数据每行包含的列比较多而行比较少的表。

表 5-2 给出了 HBase 中高表与宽表的优劣对比。

表 5-2 高表与宽表的优劣对比

	查询性能	负载均衡	元数据	事务支持
高表	√	√		
宽表			√	√

- 查询性能：由于行键是 HBase 最有效的索引，而高表行比较多，行键比较多，过滤效果更好，并且每行数据更少，因此读取的每个数据块（HBase 读取数据的最

小单元)可以包含更多的行,并且块缓存(BlockCache)或者堆外缓存(BucketCache)也可以缓存更多的行。
- 负载均衡：HBase 分片由行键区间决定,更多行可以使得分区粒度更细,分片大小更合适,负载更加均衡。
- 元数据：高表分片更多会导致元数据更多,HBase 元数据表(hbase:meta)更大,可能会给 HBase 集群主节点(HMaster)与 HBase 客户端带来更大的内存压力(客户端会缓存元数据)。
- 事务支持：HBase 只支持行级事务,因此对宽表的一行数据变更操作能够提供事务保证,而高表对多行的数据变更无事务支持。

回到用户行为日志表 s_behavior 的设计,到底是应该选用高表设计还是宽表设计呢？显然行键设计章节最后得出的表结构是一个高表。

5.4 微信朋友圈设计

2017 年 11 月 9 日,微信团队发布了《2017 年微信数据报告》,报告显示 2017 年 9 月日均登录用户超 9 亿,日发送次数达 380 亿,朋友圈视频发布次数达 6800 万,相信很多人都会对这样一个庞然大物架构的设计很感兴趣。

由于微信的体量以及爆发增长的特性,因此分布式可弹性伸缩、多机房负载均衡、容灾是系统架构与设计必须考虑的要素。HBase 就是一个很合适的存储系统,下面来一起看看如何使用 HBase 作为朋友圈存储系统的设计,同样从需求开始分析如何实现。

5.4.1 需求定义

朋友圈的核心是每个用户各自拥有的一个自己发布的相册和一个用户关注的好友的动态,称之为时间线 TimeLine。为了实现这两个功能,存储系统的设计需要考虑如下需求。
- 需要知道哪些用户关注了 LiLei 以使得这些用户能够看到 LiLei 的朋友圈。
- 需要知道 LiLei 关注了哪些用户以计算 LiLei 能够看到 Lucy 朋友圈的哪些评论。
- 为了保证用户体验,LiLei 的朋友圈打开速度要求很快。

5.4.2 问题建模

第一、二个需求可以归结为一个用户关系表的构建,该用户关系表会有两种访问模式,第一个是查询哪些用户关注了 LiLei,第二个是查询 LiLei 关注了哪些用户。第一个想到的设计可能就是从主语出发,使用用户 ID 作为行键,每列存储一个该用户关注人的用户 ID,

显然这就是一个之前提到的宽表的设计。假设该表命名为 `t_following`，表数据格式如图 5-1 所示。

t_user		
行键	列族（cf）	
	列用户名（n）	
12345	LiLei	
12346	LiLy	
12347	Lucy	
12348	HanMeiMei	
12349	Tom	

t_following				
行键	列族（cf）			
	列1	列2	列3	列4
12345	12346	12348	12349	…
12346	12348	12649		
12347	12346			

图 5-1　用户关系表-宽表

图 5-1 的两个表都包含一个列族，列族名均为 `cf`，`t_user` 表列族 `cf` 包含一列，列限定符名称为 `n`，而 `t_following` 表受益于 HBase 的无模式，列可以在使用时动态定义，那么现在有个新问题是假设 LiLei（用户 ID 为 12345）新关注了一个用户 Kate（12350），HBase 客户端在向 `t_following` 表写数据的时候需要知道写入的列限定符名称，那么此时需要定义一个规则来命名列限定符，可以使用如下两种解决方案。

- 为每个用户维护一个计数器，使用计数器得到的数字作为列限定符。该方法在取消关注某用户时需要考虑并发，为了避免并发的情况下取消关注的数据行写回相互覆盖，HBase 客户端需要查询出整行数据，遍历所有的列，找到需要删除的列限定符，然后做删除。
- 使用新关注的用户 ID，如 Kate（12350），作为列限定符。该方法的优点是客户端在插入新关注用户时无须调用服务端获取列限定符，取消关注时也无须查出整行数据，而只需直接将用户 ID 作为操作的列限定符使用。

表 `t_following` 事实上只满足了 LiLei 关注了哪些用户的需求，一个简单的 `GET` 请求即可知道 LiLei 关注了哪些用户。如果需要知道哪些用户关注了 LiLei，需要使用同样的模式新建表 `t_followed` 用来存储哪些用户关注了 LiLei。

现在已经满足了需求定义的第一个需求和第二个需求的前半部分（哪些用户关注了 LiLei），第三个需求似乎也容易满足。虽然一个 `GET` 请求即可知道 LiLei 关注了哪些用户或者哪些用户关注了 LiLei，但是这个 `GET` 请求需要读取出整行数据，然后遍历这行数据所有的列来得到 LiLei 有没有关注 Lucy，才能知道 LiLei 能不能看到 Lucy 对他们的共同关注对象 HanMeiMei 朋友圈的评论。如果 LiLei 关注了成百上千个朋友，那么注定 LiLei 的朋友圈体验没那么优雅。那么，有没有更加高效的解决方案呢？凡事都有两面性，既然我们使用宽表来设计存储架构获得了成功，那么使用高表设计又会有什么样的效果呢？

使用高表来存储用户关注列表的关键点是将列转行，可以使用"用户 ID+被关注人用户 ID"作为行键。在朋友圈消息或者评论上会显示用户的昵称，因此为了提高性能，可以把用户昵称作为单元值存储，这是一个反范式的设计，违反了数据库设计的第三范式，会导致数据的冗余。假如用户更新了昵称，那么除了更新用户表 `t_user` 之外，用户关系表 `t_following` 也需要更新，一般情况下用户更新昵称周期比较长，而且更新昵称后允许延迟展示，因此可以采用一些延迟更新之类的策略来维持数据的一致性（这种最终一致性，在过程中可能存在不一致性，也称为弱一致性）。考虑到各影响因素，我们预期用户更新昵称周期比较长，通过冗余用户昵称的代价可以提升一部分读性能，因此这里的反范式设计是一个更好的解决方案，再来看看高表设计下的数据格式，如图 5-2 所示。

同样表 `t_following` 事实上只满足了 LiLei 关注了哪些用户的需求，一个 `Get` 请求即可以知道 LiLei 是否关注了 HanMeiMei，并且该 `Get` 请求只需查询出一个单元值，查询的数据量大大小于宽表的设计，一个区间扫描即可知道 LiLei 关注了哪些用户。

如果需要知道哪些用户关注了 LiLei，需要使用同样的模式新建表 `t_followed` 用来存储哪些用户关注了 LiLei。

还有一种设计是为这种关系加一个类型，例如，`0` 代表关注，`1` 代表被关注，行键 `12345_0_12346` 表示 LiLei 关注 Lily，行键 `12346_1_12345` 表示 LiLei 被 Lily 关注，这样就可以合并 `t_following` 表与 `t_followed` 表，如图 5-3 所示。

t_user

行键	列族（cf）列用户名（n）
12345	LiLei
12346	LiLy
12347	Lucy
12348	HanMeiMei
12349	Tom

t_following

行键	列族（cf）n
12345_12346	LiLy
12345_12348	HanMeiMei
12345_12349	Tom
12346_12348	HanMeiMei
12346_12349	Tom
12347_12346	LiLy

图 5-2 用户关系表：高表

t_user

行键	列族（cf）列用户名（n）
12345	LiLei
12346	LiLy
12347	Lucy
12348	HanMeiMei
12349	Tom

t_following

行键	列族（cf）n
12345_0_12346	LiLy
12345_0_12348	HanMeiMei
12345_0_12349	Tom
12346_0_12348	HanMeiMei
12346_0_12349	Tom
12346_1_12345	LiLei
12346_1_12347	Lucy
12347_0_12346	Tom

图 5-3 用户关系表-合并关注与被关注

注意，这里行键的设计是把关系类型作为行键的第二个因子。如果把关系类型作为行键的第一个因子，那么用户关注的数据与用户被关注的数据会聚簇在一起，分别作为不同的分区分布在分区服务器。如果用户关注的数据使用比较频繁，那么负责这部分数据的分区服务器会比较繁忙，这就造成了整个集群的负载不均衡。

在线上生产环境中的大部分情况都会使用高表的设计，因为相对来说高表性能都会优于宽表，但是由于 HBase 只支持行级事务，如果某些业务要求使用事务来保证数据的强一致性，那么此时高表就可能不是一个更好的选择了。

到这里朋友圈的两个核心功能个人相册与时间线仍未实现，下面先看一看朋友圈的业务流程。

（1）HanMeiMei 发布一张朋友圈图片，该图片会上传到 HanMeiMei 最近的 CDN 缓存服务器，上传成功后返回一个图片引用地址。

（2）HanMeiMei 的微信客户端将朋友圈内容以及图片引用地址发布到微信服务器自己的相册。

（3）发布完成后如果 LiLei 刷新自己的朋友圈时间线，由于 LiLei 关注了 HanMeiMei，因此 LiLei 会看到 HanMeiMei 刚刚发布的朋友圈。

个人相册的存储表比较简单，用户发布朋友圈也只需要插入一条数据到用户相册表，行键使用用户 ID 加时间戳，列族 cf 包含两列，列 t 用来存储用户朋友圈的文字内容，列 p 用来存储用户朋友圈的图片 CDN 缓存地址，如图 5-4 所示。

t_album

行键	列族（cf）	
	t	p
12345_1521708554262	这是lilei第1条朋友圈	http://url1
12345_1521708554263	这是lilei第2条朋友圈	http://url2,http://url222
12347_1521708554262	这是Lucy的朋友圈	http://url3
12348_1521708554262	来自HanMeiMei的消息	http://url4
12349_1521708554262	Tom向你问好	http://url5

图 5-4　用户朋友圈相册表

注意到为了可读性，这里的行键设计并未做优化，为了更好地实现负载均衡以及数据的读取，对行键的用户 ID 可以做反转补齐，时间戳部分可以用 Long.MAX_VALUE-timestamp 后补齐，这样中间的连接符号"-"可以省略。

到这里由于用户的朋友圈消息已经存储下来了，用户的时间线可以根据 t_following 和 t_album 两张表计算得出。如果 LiLei 关注了 500 个用户，那么 LiLei 刷新朋友圈要扫描出这 500 个用户的朋友圈消息，然后按时间倒序排序，选取前 N 条展示到 LiLei 的朋友圈，这样的效率无疑是无法接受的，而且这 500 个用户的数据可能分布在上百台服务器，其中某些请求有很大的概率会失败，最后 LiLei 完全没法愉快地刷朋友圈了。

为了提高朋友圈用户体验，读性能必须能够保证。为了提高读性能，可以将读取的工作提前到写之后，类似于 HDFS 写多个数据副本，这里也可以使用写扩散的模型来提高读的效率，工作流程如下。

（1）HanMeiMei 发布一张朋友圈图片，该图片会上传到 HanMeiMei 最近的 CDN 缓存服务器，上传成功后返回一个图片引用地址。

（2）HanMeiMei 的微信客户端将朋友圈内容以及图片引用地址发布到微信服务器自己的相册。

（3）查找 t_followed 表找到哪些用户关注了 HanMeiMei，将 HanMeiMei 发布的朋友圈内容以及图片插入到这些用户的 TimeLine。

（4）LiLei 刷新自己的朋友圈，直接从 LiLei 的时间线表查询出数据展示。

HBase 的写性能很高，读性能相对低，时间线的设计使用写扩散模型，通过写多份冗余数据来提升读性能，因此朋友圈的发布是一个比较重的动作，但是其实对用户是无感知的，做到了对业务、用户体验无损的情况下提升性能。最后用户的时间线表设计如图 5-5 所示，采用高表设计，行键为"用户 ID+时间戳+发布者 ID"，同样为了可读性，行键并未使用反转补齐等策略做负载均衡，列族 cf 包含 3 列，列 t 用来存储用户朋友圈的文字内容，列 p 用来存储用户朋友圈的图片 CDN 缓存地址，列 u 用来存储发布者的个人基本信息（包括昵称和头像），这样用户刷新朋友圈时就无须多一次用户信息的查找。细心的读者可能会提出疑问，如果用户在同一时间发布多条朋友圈，那么行键就重复了？确实如此，但是显然同一时间，同一个用户基本不可能发布超过一条朋友圈，因此这里不做考虑。

t_timeline

行键	列族（cf）		
	t	p	u
12345_1521708554262_12348	来自HanMeiMei的消息	http://url4	HanMeiMei,http://picurlHanMeiMei
12345_1521708554263_12349	Tom向你问好	http://url5	Tom,http://picurltom
12346_1521708554262_12348	来自HanMeiMei的消息	http://url4	HanMeiMei,http://picurlHanMeiMei

图 5-5　用户时间线表设计

第 6 章

客户端 API

HBase 使用 Java 语言开发，因而 HBase 原生提供了一个 Java 语言客户端。除此之外如果想使用其他语言访问 HBase，可以使用一些能够将请求转化为 API 的代理，这些代理将 HBase Java 客户端 API 封装为与语言无关的调用协议，这样就可以使用任何语言来访问 HBase 了。HBase 本身支持了如 REST、Thrift、Avro 等代理模式。

6.1 Java 客户端使用

本章介绍 HBase Java 客户端，包括数据查询、修改、删除、建表等 API 的使用。如果项目使用 Maven 进行依赖管理，只需添加代码清单 6-1 所示的依赖即可开始使用 Java 客户端访问 HBase 集群。

代码清单 6-1　HBase Java 客户端依赖

```
<dependency>
    <groupId>org.apache.zookeeper</groupId>
    <artifactId>zookeeper</artifactId>
    <version>3.4.6</version>
</dependency>
<dependency>
    <groupId>org.apache.hbase</groupId>
    <artifactId>hbase-client</artifactId>
    <version>1.2.6</version>
</dependency>
<dependency>
    <groupId>org.apache.hbase</groupId>
    <artifactId>hbase-common</artifactId>
    <version>1.2.6</version>
```

```xml
</dependency>
<dependency>
    <groupId>org.apache.hbase</groupId>
    <artifactId>hbase-protocol</artifactId>
    <version>1.2.6</version>
</dependency>
<dependency>
    <groupId>org.apache.hbase</groupId>
    <artifactId>hbase-server</artifactId>
    <version>1.2.6</version>
</dependency>
```

HBase Java 客户端使用 Google Protocol Buffer（简称 Protobuf）与 HBase 服务端通信。Protobuf 是一种轻便高效的结构化数据存储格式，可以用于结构化数据序列化，很适合作为数据存储或数据交换格式，它具有平台无关性。HBase 1.2.6 依赖的 Protobuf 版本为 2.5.0，目前 Protobuf 的最新版本为 3.x.x，如果项目中已经使用了 3.x.x 版本的 Protobuf，要再引入 HBase 1.2.6 客户端相关依赖的 jar 包就会引起类冲突，此时可以使用 HBase 提供的 Java 客户端 shaded 依赖包，如代码清单 6-2 所示。

代码清单 6-2　HBase Java 客户端依赖 shaded 包

```xml
<dependency>
    <groupId>org.apache.hbase</groupId>
    <artifactId>hbase-shaded-client</artifactId>
    <version>1.2.6</version>
</dependency>
```

图 6-1 描述了 Java 客户端向 HBase 服务端发起请求的使用流程。

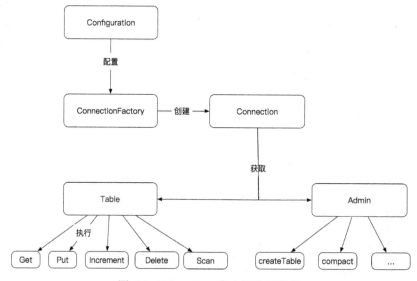

图 6-1　HBase Java 客户端使用流程

（1）构建一个 `Configuration` 实例，该实例包含了一些客户端配置项，最重要的必需的两个配置项是 HBase 集群的 ZooKeeper 地址与端口。

（2）`ConnectionFactory` 根据 `Configuration` 实例创建一个 `Connection` 对象，该 `Connection` 对象线程安全，封装了连接到一个 HBase 集群所需要的所有信息，如元数据缓存、客户端与 HMaster 和 HRegionServer 的连接等。由于创建 `Connection` 开销较大，类似于关系型数据库的连接池，因此实际使用中会将该 `Connection` 缓存起来重复使用。

（3）从 `Connection` 获取需要操作的 `Table` 的实例，该 `Table` 实例非线程安全。因为获取 `Table` 是一个轻量级操作，所以每次请求都需要获取一个新的 `Table` 实例。

（4）根据所需要做的操作类型调用 `Table` 实例的不同方法，`Get` 方法可以直接得到结果，`Scan` 操作可以得到一个 `ResultScanner` 游标，调用 `ResultScanner.next` 可以遍历 `Scan` 结果。

（5）从 `Connection` 也可以获取一个 `Admin` 实例，可以用来创建表、删除表、触发分区压缩、触发分区拆分、管理镜像和管理分区等操作。

代码清单 6-3 描述了如何获取一个 HBase 连接的过程。

代码清单 6-3　获取连接

```
1.    package com.mt.hbase.connection;
2.
3.
4.    import org.apache.hadoop.conf.Configuration;
5.    import org.apache.hadoop.hbase.HBaseConfiguration;
6.    import org.apache.hadoop.hbase.client.Connection;
7.    import org.apache.hadoop.hbase.client.ConnectionFactory;
8.
9.    import java.io.IOException;
10.
11.
12.   public class HBaseConnectionFactory {
13.
14.       private static Connection connection = null;
15.
16.       static{
17.           createConnection();
18.       }
19.
20.       private static synchronized void createConnection() {
21.
22.           Configuration configuration = HBaseConfiguration.create();
23.           configuration.set("hbase.client.pause", "100");
24.           configuration.set("hbase.client.write.buffer", "10485760");
25.           configuration.set("hbase.client.retries.number", "5");
26.           configuration.set("hbase.zookeeper.property.clientPort", "2181");
27.           configuration.set("hbase.client.scanner.timeout.period", "100000");
28.           configuration.set("hbase.rpc.timeout", "40000");
```

```
29.        configuration.set("hbase.zookeeper.quorum", "master1,master2,slave1");
30.
31.        try {
32.            connection = ConnectionFactory.createConnection(configuration);
33.        }catch(IOException ioException){
34.            throw new RuntimeException(ioException);
35.        }
36.    }
37.
38.    public static Connection getConnection() {
39.        return connection;
40.    }
41.
42.
43. }
44.
```

6.2 数据定义语言

HBase 客户端可用来创建表、删除表、修改表结构和管理分区等，应用场景包括但不局限于如下所列。

- 日志系统数据量巨大，需要分片按天存储，因而每天需要新建一个带日期的新表。
- 为了避免影响在线性能，集群关闭了自动主压缩（major compact），需要在业务非高峰期使用程序自动触发压缩。
- 为了避免影响在线性能，集群关闭了自动拆分（split）。当分区的 StoreFile 达到一定大小的时候，需要使用程序触发拆分。

接下来描述常用的几种场景，包括创建、删除表和管理分区。

6.2.1 表管理

以用户行为日志系统为例，当数据量到达一定大小后，可以考虑每天为用户行为日志表新建一个分片表，同时可以对历史数据表做清理。如代码清单 6-4 所示，该程序新建一个带当日日期后缀的表，同时删除一个月以前新建的表。

代码清单 6-4 表管理

```
1. package com.mt.hbase.chpt06.clientapi.ddl;
2.
3. import com.mt.hbase.connection.HBaseConnectionFactory;
4. import org.apache.hadoop.hbase.HColumnDescriptor;
5. import org.apache.hadoop.hbase.HTableDescriptor;
6. import org.apache.hadoop.hbase.TableName;
```

```java
7.     import org.apache.hadoop.hbase.client.Admin;
8.     import org.apache.hadoop.hbase.io.encoding.DataBlockEncoding;
9.
10.    import java.text.SimpleDateFormat;
11.    import java.util.Calendar;
12.
13.    public class TableDemo {
14.
15.        /**
16.         * 创建表
17.         * @param tableName 表名
18.         * @param familyNames 列族名
19.         * @return
20.         */
21.        public boolean createTable(String tableName, String... familyNames)
       throws Exception {
22.            Admin admin = HBaseConnectionFactory.getConnection().getAdmin();
23.            if (admin.tableExists(TableName.valueOf(tableName))) {
24.                return false;
25.            }
26.            // 通过HTableDescriptor类来描述一个表,HColumnDescriptor描述一个列族
27.            HTableDescriptor tableDescriptor = new
       HTableDescriptor(TableName.valueOf(tableName));
28.            for (String familyName : familyNames) {
29.                HColumnDescriptor oneFamily= new HColumnDescriptor(familyName);
30.                // 设置行键编码格式
31.                oneFamily.setDataBlockEncoding(DataBlockEncoding.PREFIX_TREE);
32.                // 设置数据保留多版本
33.                oneFamily.setMaxVersions(3);
34.                tableDescriptor.addFamily(oneFamily);
35.            }
36.            // 设置
37.            admin.createTable(tableDescriptor);
38.            return true;
39.        }
40.
41.        /**
42.         * 删除表
43.         * @param tableName 表名
44.         * @return
45.         */
46.        public boolean deleteTable(String tableName) throws Exception {
47.            Admin admin = HBaseConnectionFactory.getConnection().getAdmin();
48.            if (!admin.tableExists(TableName.valueOf(tableName))) {
49.                return false;
50.            }
51.            // 删除之前要将表禁用
52.            if (!admin.isTableDisabled(TableName.valueOf(tableName))) {
53.                admin.disableTable(TableName.valueOf(tableName));
54.            }
```

```
55.            admin.deleteTable(TableName.valueOf(tableName));
56.            return true;
57.        }
58.
59.
60.    public static void main(String[] args) throws Exception {
61.
62.        Calendar calendar = Calendar.getInstance();
63.        SimpleDateFormat format = new SimpleDateFormat("yyyyMMdd");
64.        TableDemo tableDemo = new TableDemo();
65.        // 新建表
66.        System.out.println(tableDemo.createTable("s_behavior"+format.format
    (calendar.getTime()),"pc","ph"));
67.
68.        calendar.add(Calendar.MONTH,-1);
69.        // 删除表
70.        System.out.println(tableDemo.deleteTable("s_behavior"+format.format
    (calendar.getTime())));
71.        }
72.
73.    }
74.
```

6.2.2 分区管理

分区的主压缩（major compact）与拆分（split）是一个比较耗时、耗资源的操作，通常在实时在线生产环境中会禁止分区的自动拆分与自动主压缩以免影响性能，因此需要在合适的时候对分区进行压缩或者拆分以免分区的 StoreFile 数量过多或者大小过大影响查询性能，代码清单 6-5 演示了如何通过 Java 客户端程序触发表的分区主压缩。

代码清单 6-5　分区主压缩

```
1.    package com.mt.hbase.chpt06.clientapi.ddl;
2.
3.    import com.mt.hbase.connection.HBaseConnectionFactory;
4.    import org.apache.hadoop.hbase.HRegionInfo;
5.    import org.apache.hadoop.hbase.HRegionLocation;
6.    import org.apache.hadoop.hbase.MetaTableAccessor;
7.    import org.apache.hadoop.hbase.TableName;
8.    import org.apache.hadoop.hbase.client.Admin;
9.    import org.apache.hadoop.hbase.protobuf.generated.AdminProtos;
10.
11.   import java.util.ArrayList;
12.   import java.util.HashMap;
13.   import java.util.List;
14.   import java.util.Map;
15.   import java.util.concurrent.Callable;
```

```java
16.    import java.util.concurrent.ExecutorService;
17.    import java.util.concurrent.Executors;
18.    import java.util.concurrent.Future;
19.
20.    public class AdminDemo {
21.
22.        private static ExecutorService executors = Executors.newFixedThreadPool(20);
23.
24.
25.        public void compact(String table) throws Exception {
26.            Admin admin = HBaseConnectionFactory.getConnection().getAdmin();
27.
28.            Map<String, List<byte[]>> serverMap = new HashMap<String, List<byte[]>>();
29.            List<HRegionInfo> regionInfos = admin.getTableRegions(TableName.valueOf(table));
30.            // 将表所有的分区按所在分区服务器分组,这样可以并发在每个分区服务器压缩一个分区
31.            for (HRegionInfo hRegionInfo : regionInfos) {
32.                HRegionLocation regionLocation = MetaTableAccessor
33.                        .getRegionLocation(HBaseConnectionFactory.getConnection(), hRegionInfo);
34.                if (serverMap.containsKey(regionLocation.getHostname())) {
35.                    serverMap.get(regionLocation.getHostname()).add(hRegionInfo.getRegionName());
36.                } else {
37.                    List<byte[]> list = new ArrayList<byte[]>();
38.                    list.add(hRegionInfo.getRegionName());
39.                    serverMap.put(regionLocation.getHostname(), list);
40.                }
41.            }
42.            List<Future<String>> futures = new ArrayList<Future<String>>();
43.            // 为每个分区服务器压缩一个分区
44.            for (Map.Entry<String, List<byte[]>> entry : serverMap.entrySet()) {
45.                futures.add(executors.submit(new HBaseCompactThread(entry)));
46.            }
47.            for (Future<String> future : futures) {
48.                System.out.println("compact results: " + future.get());
49.            }
50.        }
51.
52.
53.        class HBaseCompactThread implements Callable<String> {
54.
55.            private Map.Entry<String, List<byte[]>> entry;
56.
57.            public HBaseCompactThread(Map.Entry<String, List<byte[]>> entry) {
58.                this.entry = entry;
59.            }
60.
61.
62.            public String call() throws Exception {
63.                Admin admin = HBaseConnectionFactory.getConnection().getAdmin();
64.
65.                for (byte[] bytes : entry.getValue()) {
```

```
66.            AdminProtos.GetRegionInfoResponse.CompactionState state = admin
67.                 .getCompactionStateForRegion(bytes);
68.            // 如果分区当前状态不为主压缩，则触发主压缩
69.            if (state != AdminProtos.GetRegionInfoResponse.CompactionState.MAJOR) {
70.                admin.majorCompactRegion(bytes);
71.            }
72.            while (true) {
73.                // 休眠等待当前分区压缩结束，以免同时压缩过多分区造成分区服务器压力过大
74.                Thread.sleep(3 * 60 * 1000);
75.                state = admin.getCompactionStateForRegion(bytes);
76.
77.                if (state == AdminProtos.GetRegionInfoResponse.CompactionState.NONE) {
78.                    break;
79.                }
80.            }
81.        }
82.        return entry.getKey() + " success";
83.    }
84. }
85.
86.
87.    public static void main(String[] args) throws Exception {
88.        AdminDemo adminDemo = new AdminDemo();
89.        adminDemo.compact("s_behavior");
90.    }
91. }
```

6.3 数据操纵语言

在第 4 章介绍了用 HBase shell 来实现数据的增删改查等 DML 操作，本节介绍以 Java 客户端 API 来实现同样的功能。

6.3.1 Put

Put 用来向表中插入数据。以用户行为日志表为例，假设需要为用户 ID 12345 插入行为数据，数据部分包括行键与数据列值对，如果批量执行 Put 操作，这些 Put 请求操作的数据行键可能分布在不同的分区，并且由不同的分区服务器负责提供服务，此时客户端会将这些 Put 按分区服务器分组，再分别提交到这些分区服务器，因此一个批量 Put 操作可能会发起多个 RPC 请求，代码清单 6-6 演示了为用户 ID 12345 准备了 4 条数据批量插入用户行为日志表 s_behavior。

代码清单 6-6 Put

```java
1.  package com.mt.hbase.chpt06.clientapi.dml;
2.
3.  import com.mt.hbase.chpt05.rowkeydesign.RowKeyUtil;
4.  import com.mt.hbase.connection.HBaseConnectionFactory;
5.  import org.apache.hadoop.hbase.TableName;
6.  import org.apache.hadoop.hbase.client.HTable;
7.  import org.apache.hadoop.hbase.client.Put;
8.  import org.apache.hadoop.hbase.client.Table;
9.  import org.apache.hadoop.hbase.util.Bytes;
10.
11. import java.io.IOException;
12. import java.util.ArrayList;
13. import java.util.List;
14. import java.util.Random;
15.
16. public class PutDemo {
17.
18.     private static final String TABLE="s_behavior";
19.
20.     private static final String CF_PC="pc";
21.     private static final String CF_PHONE="ph";
22.
23.     private static final String COLUMN_VIEW="v";
24.
25.     private static final String COLUMN_ORDER="o";
26.
27.     private static final String[] ITEM_ID_ARRAY = new
    String[]{"1001","1002","1004","1009"};
28.
29.     private static final long userId = 12345;
30.
31.     private static RowKeyUtil rowKeyUtil = new RowKeyUtil();
32.
33.     public static void main(String[] args) throws IOException,
    InterruptedException {
34.         List<Put> actions = new ArrayList<Put>();
35.         Random random = new Random();
36.         for (int i = 0; i < ITEM_ID_ARRAY.length; i++) {
37.             String rowkey = generateRowkey(userId,System.currentTimeMillis(),i);
38.             Put put = new Put(Bytes.toBytes(rowkey));
39.             // 添加列
40.             put.addColumn(Bytes.toBytes(CF_PC), Bytes.toBytes(COLUMN_VIEW),
41.                 Bytes.toBytes(ITEM_ID_ARRAY[i]));
42.             if(random.nextBoolean()){
43.                 put.addColumn(Bytes.toBytes(CF_PC), Bytes.toBytes(COLUMN_ORDER),
44.                     Bytes.toBytes(ITEM_ID_ARRAY[i]));
45.             }
46.             actions.add(put);
```

```
47.        }
48.        Table table =
    HBaseConnectionFactory.getConnection().getTable(TableName.valueOf(TABLE));
49.        // 设置不启用客户端缓存，直接提交
50.        ((HTable)table).setAutoFlush(true,false);
51.
52.        // 方法一：向表写入数据
53.        table.put(actions);
54.
55.        //    Object[] results = new Object[actions.size()];
56.        // 方法二：执行 table 的批量操作，actions 可以是 Put、Delete、Get、Increment
    等操作，并且可以获取执行结果
57.        //    table.batch(actions, results);
58.
59.        // 如果启用了客户端缓存，也可以执行 flushCommits 显示提交
60.        //    ((HTable) table).flushCommits();
61.
62.    }
63.
64.    private static String generateRowkey(long userId, long timestamp, long seqId){
65.        return rowKeyUtil.formatUserId(userId)+ rowKeyUtil.formatTimeStamp(
    timestamp)+seqId;
66.    }
67. }
68.
```

6.3.2 Get

Get 请求用来根据行键从表获取一行数据。因为 HBase 行键即索引，所以 Get 的性能很快。Get 也可以指定需要获取数据的列族、列、时间范围和版本号等，如代码清单 6-7 所示。

代码清单 6-7　Get

```
1.  package com.mt.hbase.chpt06.clientapi.dml;
2.
3.  import com.mt.hbase.connection.HBaseConnectionFactory;
4.  import org.apache.hadoop.hbase.Cell;
5.  import org.apache.hadoop.hbase.CellUtil;
6.  import org.apache.hadoop.hbase.TableName;
7.  import org.apache.hadoop.hbase.client.Get;
8.  import org.apache.hadoop.hbase.client.Result;
9.  import org.apache.hadoop.hbase.client.Table;
10. import org.apache.hadoop.hbase.util.Bytes;
11.
12. import java.io.IOException;
13. import java.text.ParseException;
```

```
14.    import java.text.SimpleDateFormat;
15.    import java.util.ArrayList;
16.    import java.util.Calendar;
17.    import java.util.Date;
18.    import java.util.List;
19.
20.    public class GetDemo {
21.
22.        private static final String TABLE="s_behavior";
23.
24.        private static final String CF_PC="pc";
25.        private static final String CF_PHONE="ph";
26.
27.        private static final String COLUMN_VIEW="v";
28.
29.        private static final String COLUMN_ORDER="o";
30.
31.
32.        public static void main(String[] args) throws IOException, InterruptedException,
33.            ParseException {
34.            List<Get> gets = new ArrayList<Get>();
35.            Get oneGet = new Get(Bytes.toBytes("54321000000000000000092233705146317032071"));
36.            // 设置需要 Get 的数据列族
37.            oneGet.addFamily(Bytes.toBytes(CF_PC));
38.            // 设置需要 Get 的数据列
39.            // oneGet.addColumn(Bytes.toBytes(CF_PHONE),Bytes.toBytes(COLUMN_ORDER));
40.            // 设置 Get 的数据时间范围为 2018 年 1 月 1 日到现在
41.            String startS = "2018-01-01";
42.            SimpleDateFormat dateFormat = new SimpleDateFormat("yyyy-MM-dd");
43.            Date startDate = dateFormat.parse(startS);
44.            oneGet.setTimeRange(startDate.getTime(),System.currentTimeMillis());
45.
46.            // 设置 Get 的数据版本为 2
47.            oneGet.setMaxVersions(2);
48.            gets.add(oneGet);
49.
50.            Table table =
       HBaseConnectionFactory.getConnection().getTable(TableName.valueOf(TABLE));
51.            Result[] results = table.get(gets);
52.            for(Result result : results){
53.                if (null != result.getRow()) {
54.                    Cell[] cells = result.rawCells();
55.                    System.out.println("rowkey="+ Bytes.toString(result.getRow()));
56.                    for (Cell cell : cells) {
57.                        String qualifier = Bytes.toString(CellUtil.cloneQualifier(cell));
58.                        String value = Bytes.toString(CellUtil.cloneValue(cell));
59.                        System.out.println("qualifier="+ qualifier+",value="+value);
60.                    }
61.                }
```

```
62.         }
63.
64.        /**
65.         * 输出结果如下所示:
66.         * rowkey=5432100000000000000009223370514631703207 1
67.         * qualifier=o,value=1002
68.         * qualifier=v,value=1002
69.         */
70.    }
71. }
```

6.3.3 Scan

Scan 与 Get 都可以用来查询数据, 区别是 Scan 可以用来读取多行数据, Scan 结果的遍历类似于关系型数据库的游标。除基本的指定查询列族、列、时间范围、数据版本之外, Scan 还可以实现如下数据查询功能。

- 查询某个行键区间的数据, 结果包含开始行键, 不包含结束行键。
- 设置本次查询的事务隔离级别, 如读已提交、读未提交。
- 设置是否在服务端缓存本次扫描读取的数据块。
- 与过滤器一起使用, 在服务端过滤不需要的数据以减少传输数据量。

代码清单 6-8 演示了上述的查询功能, 过滤器的使用比较复杂, 在 6.4 节详细介绍。

代码清单 6-8   Scan

```
1.  package com.mt.hbase.chpt06.clientapi.dml;
2.
3.  import com.mt.hbase.chpt05.rowkeydesign.RowKeyUtil;
4.  import com.mt.hbase.connection.HBaseConnectionFactory;
5.  import org.apache.hadoop.hbase.Cell;
6.  import org.apache.hadoop.hbase.CellUtil;
7.  import org.apache.hadoop.hbase.TableName;
8.  import org.apache.hadoop.hbase.client.*;
9.  import org.apache.hadoop.hbase.util.Bytes;
10.
11. import java.io.IOException;
12. import java.text.ParseException;
13. import java.text.SimpleDateFormat;
14. import java.util.Date;
15.
16.
17. public class ScanDemo {
18.
19.     private static final String TABLE = "s_behavior";
20.
21.     private static final String CF_PC    = "pc";
```

6.3 数据操纵语言

```
22.      private static final String CF_PHONE = "ph";
23.
24.      private static final String COLUMN_VIEW = "v";
25.
26.      private static final String COLUMN_ORDER = "o";
27.
28.      private static final long userId = 12345;
29.
30.      public final static String MIN_TIME = "0000000000000000000";
31.      public final static String MAX_TIME = "9999999999999999999";
32.
33.      private static RowKeyUtil rowKeyUtil = new RowKeyUtil();
34.
35.
36.      public static void main(String[] args) throws IOException, InterruptedException,
37.              ParseException {
38.          Scan scan = new Scan();
39.          // 设置需要 Scan 的数据列族
40.          scan.addFamily(Bytes.toBytes(CF_PC));
41.          // 设置需要 Scan 的数据列
42.          // scan.addColumn(Bytes.toBytes(CF_PHONE),Bytes.toBytes(COLUMN_ORDER));)
43.          // 设置 small scan 以提高性能，如果扫描的数据在一个数据块内，则应该设置为 true
44.          scan.setSmall(true);
45.          // 设置扫描开始行键，结果包含开始行
46.          scan.setStartRow(Bytes.toBytes(rowKeyUtil.formatUserId(userId) + MIN_TIME));
47.          // 设置扫描结束行键，结果不包含结束行
48.          scan.setStopRow(Bytes.toBytes(rowKeyUtil.formatUserId(userId) + MAX_TIME));
49.          // 设置事务隔离级别
50.          scan.setIsolationLevel(IsolationLevel.READ_COMMITTED);
51.          // 设置每次 RPC 请求读取数据行
52.          scan.setCaching(100);
53.          // 设置 Scan 的数据时间范围为 2018 年 1 月 1 日到现在
54.          String startS = "2018-01-01";
55.          SimpleDateFormat dateFormat = new SimpleDateFormat("yyyy-MM-dd");
56.          Date startDate = dateFormat.parse(startS);
57.          scan.setTimeRange(startDate.getTime(),System.currentTimeMillis());
58.          // 设置 Scan 的数据版本为 2
59.          scan.setMaxVersions(2);
60.          // 设置是否缓存读取的数据块，如果数据会被多次读取则应该设置为 true，如果数据仅会被
             // 读取一次则应该设置为 false
61.          scan.setCacheBlocks(false);
62.
63.          Table table =
     HBaseConnectionFactory.getConnection().getTable(TableName.valueOf(TABLE));
64.          ResultScanner resultScanner = table.getScanner(scan);
65.
66.          Result result = null;
67.          while ((result = resultScanner.next()) != null) {
68.              if (result.getRow() == null) {
69.                  continue;// keyvalues=NONE
```

```
70.         }
71.         Cell[] cells = result.rawCells();
72.         System.out.println("rowkey=" + Bytes.toString(result.getRow()));
73.         for (Cell cell : cells) {
74.             String qualifier = Bytes.toString(CellUtil.cloneQualifier(cell));
75.             String value = Bytes.toString(CellUtil.cloneValue(cell));
76.             System.out.println("qualifier=" + qualifier + ",value=" + value);
77.         }
78.     }
79. }
80.
81. }
82.
```

6.3.4 Delete

Delete 用来删除数据，可以删除整行、列族、列以及某个时间戳之前的所有版本数据，使用简单，代码清单 6-9 演示了 Delete 的使用。

代码清单 6-9　Delete

```
1.  package com.mt.hbase.chpt06.clientapi.dml;
2.
3.  import com.mt.hbase.connection.HBaseConnectionFactory;
4.  import org.apache.hadoop.hbase.Cell;
5.  import org.apache.hadoop.hbase.CellUtil;
6.  import org.apache.hadoop.hbase.TableName;
7.  import org.apache.hadoop.hbase.client.Delete;
8.  import org.apache.hadoop.hbase.client.Get;
9.  import org.apache.hadoop.hbase.client.Result;
10. import org.apache.hadoop.hbase.client.Table;
11. import org.apache.hadoop.hbase.util.Bytes;
12.
13. import java.io.IOException;
14.
15. public class DeleteDemo {
16.
17.     private static final String TABLE="s_behavior";
18.
19.     private static final String CF_PC="pc";
20.     private static final String CF_PHONE="ph";
21.
22.     private static final String COLUMN_VIEW="v";
23.
24.     private static final String COLUMN_ORDER="o";
25.
26.
27.     public static void main(String[] args) throws IOException, InterruptedException {
```

```
28.         String rowkeyToDelete = "54321000000000000000092233705146317032071";
29.         Table table =
    HBaseConnectionFactory.getConnection().getTable(TableName.valueOf(TABLE));
30.
31.
32.         Get oneGet = new Get(Bytes.toBytes(rowkeyToDelete));
33.         Result result = table.get(oneGet);
34.         /**
35.          * 下行输出如下:
36.          * lineNo=1,qualifier=o,value=1002
37.          * lineNo=1,qualifier=v,value=1002
38.          */
39.         printResult(result,"1");
40.
41.         Delete deleteColumn = new Delete(Bytes.toBytes(rowkeyToDelete));
42.         // 设置需要删除的列
43.         deleteColumn.addColumn(Bytes.toBytes(CF_PC),Bytes.toBytes(COLUMN_VIEW));
44.         // 设置需要删除一天之前的数据版本
45.         deleteColumn.setTimestamp(System.currentTimeMillis() - 24*60*60*1000);
46.         table.delete(deleteColumn);
47.
48.         oneGet = new Get(Bytes.toBytes(rowkeyToDelete));
49.         result = table.get(oneGet);
50.         /**
51.          * 下行输出如下:
52.          * lineNo=2,qualifier=o,value=1002
53.          */
54.         printResult(result,"2");
55.
56.         Delete deleteFamily = new Delete(Bytes.toBytes(rowkeyToDelete));
57.         // 设置需要删除的列族
58.         deleteFamily.addFamily(Bytes.toBytes(CF_PC));
59.         table.delete(deleteFamily);
60.
61.         oneGet = new Get(Bytes.toBytes(rowkeyToDelete));
62.         result = table.get(oneGet);
63.         /**
64.          * 下行输出为空
65.          */
66.         printResult(result,"3");
67.
68.         Delete deleteRow = new Delete(Bytes.toBytes(rowkeyToDelete));
69.         // 删除整行
70.         table.delete(deleteRow);
71.
72.         oneGet = new Get(Bytes.toBytes(rowkeyToDelete));
73.         result = table.get(oneGet);
74.         /**
75.          * 下行输出为空
```

```
76.            */
77.           printResult(result,"4");
78.       }
79.
80.
81.       private static void printResult(Result result, String lineNo) {
82.           Cell[] cells = result.rawCells();
83.           for (Cell cell : cells) {
84.               String qualifier = Bytes.toString(CellUtil.cloneQualifier(cell));
85.               String value = Bytes.toString(CellUtil.cloneValue(cell));
86.               System.out.println("lineNo="+lineNo+ ", qualifier="+ qualifier+", value="+value);
87.           }
88.       }
89.   }
90.
```

6.3.5 Increment

Increment 提供了一个递增的计数器功能，该计数器的更新操作会获取行锁，因此更新是一个同步的操作，也就是说能够保证计数器递增的原子性。但是因为 Get 操作不会获取该行锁，所以 Increment 的结果可能不会马上被读取，这就是我们常说的在过程中的弱一致性，但能够保证最终结果的强一致性，代码清单 6-10 演示了这一过程。

代码清单 6-10　Increment

```
1.    package com.mt.hbase.chpt06.clientapi.dml;
2.
3.    import com.mt.hbase.connection.HBaseConnectionFactory;
4.    import org.apache.hadoop.hbase.Cell;
5.    import org.apache.hadoop.hbase.CellUtil;
6.    import org.apache.hadoop.hbase.TableName;
7.    import org.apache.hadoop.hbase.client.Get;
8.    import org.apache.hadoop.hbase.client.Increment;
9.    import org.apache.hadoop.hbase.client.Result;
10.   import org.apache.hadoop.hbase.client.Table;
11.   import org.apache.hadoop.hbase.util.Bytes;
12.
13.   import java.util.ArrayList;
14.   import java.util.List;
15.
16.   public class IncrDemo {
17.
18.       private static final String TABLE = "s_behavior";
19.
20.       private static final String CF_PC = "pc";
21.
```

```
22.     private static final String COLUMN_FOR_INCR = "i";
23.
24.     public static void main(String[] args) throws InterruptedException {
25.
26.
27.         List<Thread> incrThreadList = new ArrayList<Thread>();
28.         for(int i=0; i < 20; i++){
29.             IncrThread incrThread = new IncrThread();
30.             Thread t = new Thread(incrThread);
31.             t.setName("Thread_"+ i);
32.             incrThreadList.add(t);
33.             t.start();
34.         }
35.
36.         for(Thread incrThread: incrThreadList){
37.             incrThread.join();
38.         }
39.
40.     }
41.
42.     static class IncrThread implements Runnable{
43.
44.         @Override public void run() {
45.             try {
46.                 Table table = HBaseConnectionFactory.getConnection().getTable(TableName.valueOf(TABLE));
47.                 Increment increment = new Increment(Bytes.toBytes("rowkeyforincr"));
48.                 increment.addColumn(Bytes.toBytes(CF_PC), Bytes.toBytes(COLUMN_FOR_INCR), 1);
49.                 table.increment(increment);
50.
51.                 Get oneGet = new Get(Bytes.toBytes("rowkeyforincr"));
52.                 Result getResult = table.get(oneGet);
53.                 Cell[] getCells = getResult.rawCells();
54.                 for (Cell cell : getCells) {
55.                     String qualifier = Bytes.toString(CellUtil.cloneQualifier(cell));
56.                     Long value = Bytes.toLong(CellUtil.cloneValue(cell));
57.                     System.out.println(Thread.currentThread().getName()+": qualifier=" + qualifier + ",value=" + value);
58.                 }
59.
60.             }catch(Exception e){
61.                 System.out.println("incr failed"+ e.getMessage());
62.             }
63.         }
64.     }
65. }
66.
```

代码清单 6-10 启动 20 个进程分别对 s_behavior 表的行 rowkeyforincr 做递增 1 的操作，递增完成后马上对行 rowkeyforincr 做查询，代码清单 6-11 列出了查询的结果，该结果验证了过程中的弱一致性与最终的强一致性。

代码清单 6-11　Increment 结果

```
Thread_14: qualifier=i,value=14
Thread_18: qualifier=i,value=15
Thread_0: qualifier=i,value=14
Thread_6: qualifier=i,value=15
Thread_9: qualifier=i,value=15
Thread_5: qualifier=i,value=15
Thread_16: qualifier=i,value=15
Thread_2: qualifier=i,value=15
Thread_7: qualifier=i,value=15
Thread_17: qualifier=i,value=15
Thread_15: qualifier=i,value=15
Thread_4: qualifier=i,value=15
Thread_10: qualifier=i,value=15
Thread_12: qualifier=i,value=15
Thread_1: qualifier=i,value=18
Thread_8: qualifier=i,value=19
Thread_11: qualifier=i,value=20
Thread_19: qualifier=i,value=20
Thread_13: qualifier=i,value=20
Thread_3: qualifier=i,value=20
```

6.4　过滤器

过滤器（filter）为 HBase 提供了与关系型数据库一样支持复杂数据查询的能力，但是如果过滤器使用不当或者过于复杂可能导致性能的下降，因此复杂的过滤器组合使用必须经过性能测试才能应用到生产环境当中。

6.4.1　过滤器简介

过滤器根据行键、列族、列、时间戳等条件来对数据进行过滤，这些过滤器根据作用对象的不同而具备不同的过滤效率，行键过滤效率最高，某些过滤器（如 PageFilter）支持条件满足时能够提前停止扫描以优化性能。

过滤器在服务端与 Scan、Get 一起使用来过滤掉不需要的数据以减少在服务端与客户端传输的数据量，所有的过滤器均继承自抽象类 org.apache.hadoop.hbase.filter.Filter.java，图 6-2 列出了常用的一些过滤器类图。

6.4 过滤器

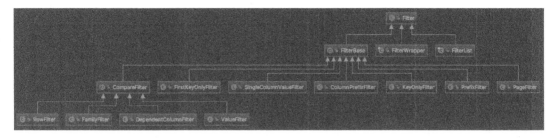

图 6-2 常用过滤器类图

Filter.java 类描述了过滤器各个方法的执行顺序，通过按顺序执行这些方法，过滤器可以过滤掉不需要的行、列族、列、转换或者直接修改最终的数据，这些方法的执行顺序以及作用如图 6-3 所示。

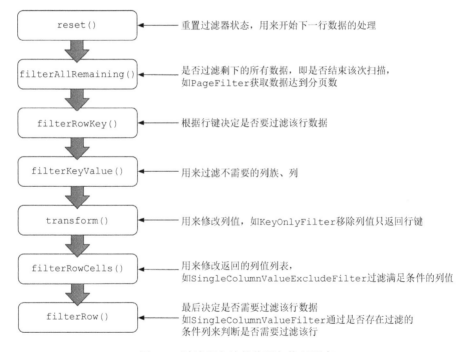

图 6-3 过滤器方法的作用与执行顺序

不同过滤器之间不一定兼容，有的不能和 Scan 一起使用，有的不能和 Get 一起使用，甚至不同过滤器添加到过滤器列表的顺序也影响扫描结果（如 KeyOnlyFilter 与多个 SingleColumnValueFilter 一起添加到 FilterList 使用，这是 FilterList 的一个 bug，在 HBase 2.x 版本，FilterList 有个重构来修复这个问题，被分为 And 和 Or 两个分支），表 6-1 列出了不同过滤器之间的兼容性。

表6-1 常用过滤器使用兼容性

过滤器	Batch	Skip	WhileMatch	FilterList	Get	Scan
RowFilter	支持	支持	支持	支持	不支持	支持
FamilyFilter	支持	支持	支持	支持	支持	支持
QualifierFilter	支持	支持	支持	支持	支持	支持
ValueFilter	支持	支持	支持	支持	支持	支持
SingleColumnValueFilter	支持	支持	支持	支持	不支持	支持
SingleColumnValueExcludeFilter	支持	支持	支持	支持	不支持	支持
PrefixFilter	支持	不支持	支持	支持	不支持	支持
PageFilter	支持	不支持	支持	支持	不支持	支持
KeyOnlyFilter	支持	支持	支持	支持	支持	支持
FirstKeyOnlyFilter	支持	支持	支持	支持	支持	支持
InclusiveStopFilter	支持	不支持	支持	支持	不支持	支持
TimestampsFilter	支持	支持	支持	支持	支持	支持
SkipFilter	支持	条件	条件	支持	不支持	支持
WhileMatchFilter	支持	条件	条件	支持	不支持	支持
FilterList	支持	条件	条件	支持	支持	支持
DependentColumnFilter	不支持	支持	支持	支持	支持	支持
ColumnCountGetFilter	支持	支持	支持	支持	支持	支持
ColumnPaginationFilter	支持	支持	支持	支持	支持	支持
ColumnPrefixFilter	支持	支持	支持	支持	支持	支持
ColumnRangeFilter	支持	支持	支持	支持	支持	支持

表6-1中各列的含义说明如下。

- `Batch`：过滤器是否支持 `Scan.batch` 方法使用，`Scan.batch` 方法可能导致一行数据分批返回，而 `DependentColumnFilter` 因为需要整行的数据列来过滤，所以不支持 `Batch`。
- `Skip`：过滤器是否可以被 `SkipFilter` 包装使用，"条件"表示这些过滤器本身是装饰者或者包装者类型，依赖其包装的过滤器而决定能不能被 `SkipFilter` 进一步包装使用。
- `WhileMatch`：过滤器是否可以被 `WhileMatchFilter` 包装使用。
- `FilterList`：过滤器是否可以被 `FilterList` 包装使用。
- `Get`：过滤器是否可以被 `Get` 实例使用并生效。
- `Scan`：过滤器是否可以被 `Scan` 实例使用并生效。

比较类型的过滤器构造函数通常包括两个类型的参数。

- 操作符：CompareOp 列出了操作符的枚举，包括 LESS、LESS_OR_EQUAL、EQUAL、NOT_EQUAL、GREATER_OR_EQUAL、GREATER 和 NO_OP。
- 比较器：因为 HBase 存储的数据格式为字节码，所以所有的比较器都继承自 ByteArrayComparable。常用的比较器包括 SubStringComparator、RegexStringComparator 和 BinaryComparator 等，如果有需要也可以定义自己的比较器。

6.4.2 过滤器使用

接下来介绍一些常用过滤器的使用方式，相关示例程序的输出结果均基于代码清单 6-12 所示的表数据。

代码清单 6-12　s_behavior 表数据

```
hbase(main):002:0> scan 's_behavior'
ROW                                            COLUMN+CELL
 543210000000000000000092233705133122008861 column=pc:v, timestamp=1523541994550,
 value=1002
 543210000000000000000092233705133122008862 column=pc:o, timestamp=1523541994550,
 value=1004
 543210000000000000000092233705133122008862 column=pc:v, timestamp=1523541994550,
 value=1004
 543210000000000000000092233705133122008863 column=pc:o, timestamp=1523541994550,
 value=1009
 543210000000000000000092233705133122008863 column=pc:v, timestamp=1523541994550,
 value=1009
 543210000000000000000092233705133122009210 column=pc:o, timestamp=1523541994550,
 value=1001
 543210000000000000000092233705133122009210 column=pc:v, timestamp=1523541994550,
 value=1001
4 row(s) in 0.1810 seconds
```

为了代码的简洁与可重用，提取了一部分公用代码到类 BaseDemo.java，包括一些表相关常量的定义以及查询结果的打印输出，之后的过滤器使用介绍示例代码类均继承了 BaseDemo 类，BaseDemo 类的定义如代码清单 6-13 所示。

代码清单 6-13　BaseDemo.java

```
1.    package com.mt.hbase.chpt06.clientapi;
2.
3.    import org.apache.hadoop.hbase.Cell;
4.    import org.apache.hadoop.hbase.CellUtil;
5.    import org.apache.hadoop.hbase.client.Result;
6.    import org.apache.hadoop.hbase.client.ResultScanner;
7.    import org.apache.hadoop.hbase.util.Bytes;
```

```
8.
9.    import java.io.IOException;
10.
11.   public class BaseDemo {
12.
13.       protected static final String TABLE = "s_behavior";
14.
15.       protected static final String CF_PC="pc";
16.       protected static final String CF_PHONE="ph";
17.
18.       protected static final String COLUMN_VIEW="v";
19.
20.       protected static final String COLUMN_ORDER="o";
21.
22.
23.       protected static void printResult(ResultScanner resultScanner) throws IOException {
24.           Result result = null;
25.           while ((result = resultScanner.next()) != null) {
26.               if (result.getRow() == null) {
27.                   continue;// keyvalues=NONE
28.               }
29.               Cell[] cells = result.rawCells();
30.               System.out.println("rowkey=" + Bytes.toString(result.getRow()));
31.               for (Cell cell : cells) {
32.                   String qualifier = Bytes.toString(CellUtil.cloneQualifier(cell));
33.                   String value = Bytes.toString(CellUtil.cloneValue(cell));
34.                   System.out.println("qualifier=" + qualifier + ",value=" + value);
35.               }
36.           }
37.       }
38.   }
39.
```

1. KeyOnlyFilter

该过滤器使得查询结果只返回行键，主要是重写了 transformCell 方法，将单元格的值全部替换为空，使用方式如代码清单 6-14 所示。

代码清单 6-14 KeyOnlyFilter 使用

```
1.    package com.mt.hbase.chpt06.clientapi.filter;
2.
3.    import com.mt.hbase.chpt06.clientapi.BaseDemo;
4.    import com.mt.hbase.connection.HBaseConnectionFactory;
5.    import org.apache.hadoop.hbase.TableName;
6.    import org.apache.hadoop.hbase.client.ResultScanner;
7.    import org.apache.hadoop.hbase.client.Scan;
8.    import org.apache.hadoop.hbase.client.Table;
9.    import org.apache.hadoop.hbase.filter.KeyOnlyFilter;
10.
```

```
11.    public class KeyOnlyFilterDemo extends BaseDemo {
12.
13.
14.        public static void main(String[] args) throws Exception {
15.
16.            Scan scan = new Scan();
17.
18.            /**
19.             * 查询表 s_behavior 所有数据的行键
20.             */
21.            scan.setFilter(new KeyOnlyFilter());
22.
23.            Table table =
       HBaseConnectionFactory.getConnection().getTable(TableName.valueOf(TABLE));
24.            ResultScanner resultScanner = table.getScanner(scan);
25.
26.            printResult(resultScanner);
27.            /**
28.             输出结果:
29.             rowkey=5432100000000000000009223370513312200886l
30.             qualifier=v,value=
31.             rowkey=54321000000000000000092233705133122008862
32.             qualifier=o,value=
33.             qualifier=v,value=
34.             rowkey=54321000000000000000092233705133122008863
35.             qualifier=o,value=
36.             qualifier=v,value=
37.             rowkey=54321000000000000000092233705133122009210
38.             qualifier=o,value=
39.             qualifier=v,value=
40.             */
41.        }
42.
43.
44.    }
45.
```

2. FirstKeyOnlyFilter

该过滤器使得查询结果只返回每行的第一个单元值，它通常与 `KeyOnlyFilter` 一起使用来执行高效的行统计操作，其用法如代码清单 6-15 所示。

代码清单 6-15　FirstKeyOnlyFilter 使用

```
1.  package com.mt.hbase.chpt06.clientapi.filter;
2.
3.  import com.mt.hbase.chpt06.clientapi.BaseDemo;
4.  import com.mt.hbase.connection.HBaseConnectionFactory;
5.  import org.apache.hadoop.hbase.TableName;
6.  import org.apache.hadoop.hbase.client.ResultScanner;
```

```
7.   import org.apache.hadoop.hbase.client.Scan;
8.   import org.apache.hadoop.hbase.client.Table;
9.   import org.apache.hadoop.hbase.filter.FirstKeyOnlyFilter;
10.
11.  public class FirstKeyOnlyFilterDemo extends BaseDemo {
12.
13.
14.      private static final String TABLE = "s_behavior";
15.
16.      public static void main(String[] args) throws Exception {
17.
18.          Scan scan = new Scan();
19.
20.          /**
21.           * 查询表 s_behavior 所有数据,每行只返回第一列,通常与 KeyOnlyFilter 一起使用
22.           */
23.          scan.setFilter(new FirstKeyOnlyFilter());
24.
25.          Table table =
     HBaseConnectionFactory.getConnection().getTable(TableName.valueOf(TABLE));
26.          ResultScanner resultScanner = table.getScanner(scan);
27.
28.          printResult(resultScanner);
29.          /**
30.           输出结果:
31.           rowkey=5432100000000000000092233705133122008861
32.           qualifier=v,value=1002
33.           rowkey=5432100000000000000092233705133122008862
34.           qualifier=o,value=1004
35.           rowkey=5432100000000000000092233705133122008863
36.           qualifier=o,value=1009
37.           rowkey=5432100000000000000092233705133122009210
38.           qualifier=o,value=1001
39.           */
40.      }
41.
42.  }
43.
```

3. PrefixFilter

该过滤器用来匹配行键包含指定前缀的数据。这个过滤器所实现的功能其实也可以由 RowFilter 结合 RegexComparator 或者 Scan 指定开始和结束行键来实现,代码清单 6-16 使用 PrefixFilter 实现了查询用户 ID 为 12345 所有的用户行为数据。

代码清单 6-16　PrefixFilter 使用

```
1.   package com.mt.hbase.chpt06.clientapi.filter;
2.
```

```
3.    import com.mt.hbase.chpt05.rowkeydesign.RowKeyUtil;
4.    import com.mt.hbase.chpt06.clientapi.BaseDemo;
5.    import com.mt.hbase.connection.HBaseConnectionFactory;
6.    import org.apache.hadoop.hbase.TableName;
7.    import org.apache.hadoop.hbase.client.ResultScanner;
8.    import org.apache.hadoop.hbase.client.Scan;
9.    import org.apache.hadoop.hbase.client.Table;
10.   import org.apache.hadoop.hbase.filter.PrefixFilter;
11.   import org.apache.hadoop.hbase.util.Bytes;
12.
13.   public class PrefixFilterDemo extends BaseDemo {
14.
15.
16.       private static final String TABLE = "s_behavior";
17.
18.       public static void main(String[] args) throws Exception {
19.           Scan scan = new Scan();
20.           RowKeyUtil rowKeyUtil = new RowKeyUtil();
21.
22.           /**
23.            * 查询用户 12345 的所有数据
24.            */
25.           PrefixFilter prefixFilter =
     new PrefixFilter(Bytes.toBytes(rowKeyUtil.formatUserId(12345)));
26.           scan.setFilter(prefixFilter);
27.
28.           Table table =
     HBaseConnectionFactory.getConnection().getTable(TableName.valueOf(TABLE));
29.           ResultScanner resultScanner = table.getScanner(scan);
30.
31.           printResult(resultScanner);
32.       }
33.
34.   }
35.
```

4. RowFilter

该过滤器通过行键来匹配满足条件的数据行。例如，使用 `BinaryComparator` 可以查询出具有某个行键的数据行、通过比较运算符（如示例代码使用 `CompareFilter.CompareOp.LESS`）筛选出符合某一条件的多条数据。代码清单 6-17 使用 `RowFilter` 实现了查询用户（ID 为 12345）2018 年 1 月 1 日之后产生的行为数据（注意，这里使用了行键自然排序的特性，`RowKeyUtil` 对时间戳做了优化使得数据越新越排在前面）。

代码清单 6-17　RowFilter 使用

```
1.    package com.mt.hbase.chpt06.clientapi.filter;
2.
3.    import com.mt.hbase.chpt05.rowkeydesign.RowKeyUtil;
```

```java
4.   import com.mt.hbase.chpt06.clientapi.BaseDemo;
5.   import com.mt.hbase.connection.HBaseConnectionFactory;
6.   import org.apache.hadoop.hbase.TableName;
7.   import org.apache.hadoop.hbase.client.ResultScanner;
8.   import org.apache.hadoop.hbase.client.Scan;
9.   import org.apache.hadoop.hbase.client.Table;
10.  import org.apache.hadoop.hbase.filter.BinaryComparator;
11.  import org.apache.hadoop.hbase.filter.CompareFilter;
12.  import org.apache.hadoop.hbase.filter.RowFilter;
13.  import org.apache.hadoop.hbase.util.Bytes;
14.
15.  import java.text.SimpleDateFormat;
16.  import java.util.Date;
17.
18.  public class RowFilterDemo extends BaseDemo {
19.
20.      public static void main(String[] args) throws Exception {
21.
22.          RowKeyUtil rowKeyUtil = new RowKeyUtil();
23.          Scan scan = new Scan();
24.          /**
25.           * 查询用户 ID 为 12345 的用户 2018 年 1 月 1 日之后的数据
26.           */
27.          String startS = "2018-01-01";
28.          SimpleDateFormat dateFormat = new SimpleDateFormat("yyyy-MM-dd");
29.          Date startDate = dateFormat.parse(startS);
30.
31.          String rowKey =
     rowKeyUtil.formatUserId(12345)+rowKeyUtil.formatTimeStamp(startDate.getTime());
32.
33.          RowFilter rowFilter = new RowFilter(CompareFilter.CompareOp.LESS, new
     BinaryComparator(
34.                  Bytes.toBytes(rowKey)));
35.          scan.setFilter(rowFilter);
36.
37.          Table table =
     HBaseConnectionFactory.getConnection().getTable(TableName.valueOf(TABLE));
38.          ResultScanner resultScanner = table.getScanner(scan);
39.
40.          printResult(resultScanner);
41.          /**
42.           * 输出结果:
43.           * rowkey=543210000000000000009223370513312200886|
44.           * qualifier=v,value=1002
45.           * rowkey=543210000000000000009223370513312200886|
46.           * qualifier=o,value=1004
47.           * qualifier=v,value=1004
48.           * rowkey=543210000000000000009223370513312200886|
49.           * qualifier=o,value=1009
50.           * qualifier=v,value=1009
```

```
51.            rowkey=54321000000000000000092233705133122009210
52.            qualifier=o,value=1001
53.            qualifier=v,value=1001
54.            */
55.        }
56.
57.    }
```

5. SingleColumnValueFilter

该过滤器类似于关系型数据库的 where 条件语句,通过判断数据行指定的列限定符对应的值是否匹配指定的条件(如等于 1001)来决定是否将该数据行返回。由于 HBase 支持动态模式,因此有些数据行可能不包含 SingleColumnValueFilter 指定的列限定符,代码 singleColumnValueFilter.setFilterIfMissing(true);可以用来决定查询结果是否返回这些不包含指定列限定符的数据行,默认 false 表示返回,true 表示不返回,如代码清单 6-18 所示。

代码清单 6-18 SingleColumnValueFilter 使用

```
1.   package com.mt.hbase.chpt06.clientapi.filter;
2.
3.   import com.mt.hbase.chpt06.clientapi.BaseDemo;
4.   import com.mt.hbase.connection.HBaseConnectionFactory;
5.   import org.apache.hadoop.hbase.TableName;
6.   import org.apache.hadoop.hbase.client.ResultScanner;
7.   import org.apache.hadoop.hbase.client.Scan;
8.   import org.apache.hadoop.hbase.client.Table;
9.   import org.apache.hadoop.hbase.filter.CompareFilter;
10.  import org.apache.hadoop.hbase.filter.SingleColumnValueFilter;
11.  import org.apache.hadoop.hbase.util.Bytes;
12.
13.  public class SingleColumnValueFilterDemo extends BaseDemo {
14.
15.      private static final String TABLE = "s_behavior";
16.
17.      private static final String CF_PC="pc";
18.      private static final String CF_PHONE="ph";
19.
20.      private static final String COLUMN_VIEW="v";
21.
22.      private static final String COLUMN_ORDER="o";
23.
24.      public static void main(String[] args) throws Exception {
25.
26.          Scan scan = new Scan();
27.          /**
28.           * 查询表 s_behavior 列限定符'o',值为 1004 的数据
29.           */
```

```
30.         SingleColumnValueFilter singleColumnValueFilter = new
    SingleColumnValueFilter(Bytes.toBytes(CF_PC), Bytes.toBytes(COLUMN_ORDER),
31.             CompareFilter.CompareOp.EQUAL, Bytes.toBytes("1004"));
32.         /**
33.          * 如果数据行不包含列限定符'o' 则不返回该行
34.          * 如果设置为 false，则会返回不包含列限定符'o'的数据行
35.          */
36.         singleColumnValueFilter.setFilterIfMissing(true);
37.
38.         scan.setFilter(singleColumnValueFilter);
39.         Table table =
    HBaseConnectionFactory.getConnection().getTable(TableName.valueOf(TABLE));
40.         ResultScanner resultScanner = table.getScanner(scan);
41.
42.         printResult(resultScanner);
43.         /**
44.          输出结果：
45.          rowkey=5432100000000000000092233705133122008862
46.          qualifier=o,value=1004
47.          qualifier=v,value=1004
48.          */
49.     }
50. }
51.
```

6. TimestampsFilter

该过滤器可以用来过滤某个时间戳的数据，可以与 `Get` 和 `Scan` 一起使用。同时 `Get` 和 `Scan` 提供的 `setTimeStamp` 和 `setTimeRange` 方法可以实现同样的功能，如代码清单 6-19 所示。

代码清单 6-19　TimestampsFilter 使用

```
1.  package com.mt.hbase.chpt06.clientapi.filter;
2.
3.  import com.mt.hbase.chpt06.clientapi.BaseDemo;
4.  import com.mt.hbase.connection.HBaseConnectionFactory;
5.  import org.apache.hadoop.hbase.TableName;
6.  import org.apache.hadoop.hbase.client.ResultScanner;
7.  import org.apache.hadoop.hbase.client.Scan;
8.  import org.apache.hadoop.hbase.client.Table;
9.  import org.apache.hadoop.hbase.filter.TimestampsFilter;
10.
11. import java.util.ArrayList;
12. import java.util.List;
13.
14. public class TimestampsFilterDemo extends BaseDemo {
15.
16.
```

```java
17.    public static void main(String[] args) throws Exception {
18.
19.        Scan scan = new Scan();
20.
21.        /**
22.         * 查询表 s_behavior hbase 时间戳为 1523541994550 的数据
23.         */
24.        List<Long> timeStampList = new ArrayList<Long>();
25.        timeStampList.add(1523541994550L);
26.        TimestampsFilter timestampsFilter = new TimestampsFilter(timeStampList);
27.
28.        scan.setFilter(timestampsFilter);
29.
30.        /**
31.         也可以使用如下命令替换:
32.         scan.setTimeStamp(1523541994550L);
33.         scan.setTimeRange(1523541994550L,1524550412164L);
34.         */
35.
36.        Table table =
       HBaseConnectionFactory.getConnection().getTable(TableName.valueOf(TABLE));
37.        ResultScanner resultScanner = table.getScanner(scan);
38.
39.        printResult(resultScanner);
40.        /**
41.         输出结果:
42.         rowkey=5432100000000000000009223370513312200886I
43.         qualifier=v,value=1002
44.         rowkey=54321000000000000000092233705133122008862
45.         qualifier=o,value=1004
46.         qualifier=v,value=1004
47.         rowkey=54321000000000000000092233705133122008863
48.         qualifier=o,value=1009
49.         qualifier=v,value=1009
50.         rowkey=54321000000000000000092233705133122009210
51.         qualifier=o,value=1001
52.         qualifier=v,value=1001
53.         */
54.    }
55.  }
```

7. ValueFilter

该过滤器使用单元格的值来过滤数据，只有满足指定条件、指定值的单元格才会被返回。与 `SingleColumnValueFilter` 不同的是后者需要指定匹配的列限定符并且返回的是整行数据。代码清单 6-20 使用 `ValueFilter` 查询出了用户行为记录表中商品 1001 的浏览数据和下单数据，这里使用了 `BinaryComparator` 比较器使得只有值等于 1001 的单元格才会被返回。

代码清单 6-20　ValueFilter 使用

```
1.   package com.mt.hbase.chpt06.clientapi.filter;
2.   
3.   import com.mt.hbase.chpt06.clientapi.BaseDemo;
4.   import com.mt.hbase.connection.HBaseConnectionFactory;
5.   import org.apache.hadoop.hbase.TableName;
6.   import org.apache.hadoop.hbase.client.ResultScanner;
7.   import org.apache.hadoop.hbase.client.Scan;
8.   import org.apache.hadoop.hbase.client.Table;
9.   import org.apache.hadoop.hbase.filter.BinaryComparator;
10.  import org.apache.hadoop.hbase.filter.CompareFilter;
11.  import org.apache.hadoop.hbase.filter.ValueFilter;
12.  import org.apache.hadoop.hbase.util.Bytes;
13.  
14.  public class ValueFilterDemo extends BaseDemo {
15.  
16.      private static final String TABLE = "s_behavior";
17.  
18.      public static void main(String[] args) throws Exception {
19.  
20.          Scan scan = new Scan();
21.  
22.          /**
23.           * 查询表 s_behavior 包含商品 1001 记录的数据
24.           * 注意如果一行有多列数据,只有值等于 1001 数据的一列会被返回
25.           */
26.          ValueFilter valueFilter = new ValueFilter(CompareFilter.CompareOp.EQUAL, new BinaryComparator(
27.                  Bytes.toBytes("1001")));
28.          scan.setFilter(valueFilter);
29.  
30.          Table table = HBaseConnectionFactory.getConnection().getTable(TableName.valueOf(TABLE));
31.          ResultScanner resultScanner = table.getScanner(scan);
32.  
33.          printResult(resultScanner);
34.          /**
35.           * 输出结果:
36.           * rowkey=5432100000000000000092233705133122009210
37.           * qualifier=o,value=1001
38.           * qualifier=v,value=1001
39.           */
40.      }
41.  }
```

8. WhileMatchFilter

该过滤器与 `SkipFilter` 同属于装饰或者包装器类型过滤器。两者的区别是当 `WhileMatchFilter` 包装的过滤器条件不被满足时, `WhileMatchFilter` 即会停止往下

扫描,返回已经扫描到的数据,而 `SkipFilter` 则是当包装的过滤器条件满足时跳过该满足条件的数据行。代码清单 6-21 使用 `WhileMatchFilter` 包装 `ValueFilter` 用来查询匹配到包含 1009 单元格之前的所有数据。

代码清单 6-21　ValueFilter 使用

```java
1.  package com.mt.hbase.chpt06.clientapi.filter;
2.
3.  import com.mt.hbase.chpt06.clientapi.BaseDemo;
4.  import com.mt.hbase.connection.HBaseConnectionFactory;
5.  import org.apache.hadoop.hbase.TableName;
6.  import org.apache.hadoop.hbase.client.ResultScanner;
7.  import org.apache.hadoop.hbase.client.Scan;
8.  import org.apache.hadoop.hbase.client.Table;
9.  import org.apache.hadoop.hbase.filter.*;
10. import org.apache.hadoop.hbase.util.Bytes;
11.
12. public class WhileMatchFilterDemo extends BaseDemo {
13.
14.     public static void main(String[] args) throws Exception {
15.
16.         Scan scan = new Scan();
17.         /**
18.          * 查询表 s_behavior
19.          * 相当于 while 执行,当条件不满足了即返回结果
20.          * 如下过滤条件相当于遇到 1009 的数据内容就返回
21.          */
22.         WhileMatchFilter whileMatchFilter = new WhileMatchFilter(new ValueFilter(CompareFilter.CompareOp.NOT_EQUAL,
23.             new BinaryComparator(Bytes.toBytes("1009"))));
24.
25.         scan.setFilter(whileMatchFilter);
26.         Table table = HBaseConnectionFactory.getConnection().getTable(TableName.valueOf(TABLE));
27.         ResultScanner resultScanner = table.getScanner(scan);
28.
29.         printResult(resultScanner);
30.         /**
31.          * 输出结果:
32.          * rowkey=54321000000000000000009223370513312200886 1
33.          * qualifier=v,value=1002
34.          * rowkey=54321000000000000000009223370513312200886 2
35.          * qualifier=o,value=1004
36.          * qualifier=v,value=1004
37.          */
38.     }
39. }
40.
```

9. FilterList

FilterList 也是一个包装类型的过滤器，可以用来包装一个有序的 Filter 列表。因为 FilterList 也继承自 Filter，所以可以通过这个特性来构造一个有层次结构的嵌套的 FilterList。

FilterList 的构造函数涉及两个参数，一个是包装的过滤器列表，另一个是列表的操作关系，包括 FilterList.Operator.MUST_PASS_ALL 和 FilterList.Operator.MUST_PASS_ONE，前者表示查询的结果数据行需要满足包装的过滤器列表的所有过滤条件，后者表示只需要满足包装的过滤器列表中任意一个过滤器的过滤条件。因此，对于操作关系 MUST_PASS_ALL，只要发现某个过滤器过滤条件不被满足，查询扫描即会停止返回，而 MUST_PASS_ONE 则需要对过滤器列表里面的过滤器全部执行过滤流程，MUST_PASS_ALL 效率稍高。

代码清单 6-22 演示了 FilterList 的使用，实现了两次查询，第二个查询使用了嵌套的 FilterList 来实现查询商品 ID 等于 1002 或者 1004 的相关数据结果。如果仔细看第二个查询的结果，会发现其实 KeyOnlyFilter 没起作用，这就是前面提到的 FilterList 执行顺序的 bug。

代码清单 6-22　FilterList 使用 1

```
1.    package com.mt.hbase.chpt06.clientapi.filter;
2.
3.    import com.mt.hbase.chpt06.clientapi.BaseDemo;
4.    import com.mt.hbase.connection.HBaseConnectionFactory;
5.    import org.apache.hadoop.hbase.TableName;
6.    import org.apache.hadoop.hbase.client.ResultScanner;
7.    import org.apache.hadoop.hbase.client.Scan;
8.    import org.apache.hadoop.hbase.client.Table;
9.    import org.apache.hadoop.hbase.filter.*;
10.   import org.apache.hadoop.hbase.util.Bytes;
11.
12.   import java.util.ArrayList;
13.   import java.util.List;
14.
15.   public class FilterListDemo extends BaseDemo {
16.
17.       public static void main(String[] args) throws Exception {
18.
19.           Table table = 
      HBaseConnectionFactory.getConnection().getTable(TableName.valueOf(TABLE));
20.
21.           /**
22.            * 查询表 s_behavior 满足如下条件的数据行
23.            * 1.列限定符 pc:o 值等于 1004
24.            * 2.列数据时间戳等于 1523541994550L
```

```
25.         * 3.只返回数据的行键，不返回列数据值
26.         */
27.        List<Filter> filters = new ArrayList<Filter>();
28.        SingleColumnValueFilter singleColumnValueFilter = new
    SingleColumnValueFilter(Bytes.toBytes(CF_PC), Bytes.toBytes(COLUMN_ORDER),
29.            CompareFilter.CompareOp.EQUAL, Bytes.toBytes("1004"));
30.        List<Long> timeStampList = new ArrayList<Long>();
31.        timeStampList.add(1523541994550L);
32.        TimestampsFilter timestampsFilter = new TimestampsFilter(timeStampList);
33.        filters.add(singleColumnValueFilter);
34.        filters.add(timestampsFilter);
35.        filters.add(new KeyOnlyFilter());
36.        FilterList filterList = new FilterList(FilterList.Operator.MUST_PASS_ALL, filters);
37.        Scan scanOnlyKey = new Scan();
38.        scanOnlyKey.setFilter(filterList);
39.        ResultScanner resultScanner = table.getScanner(scanOnlyKey);
40.        printResult(resultScanner);
41.        /**
42.         输出结果：
43.         rowkey=5432100000000000000009223370513312200862
44.         qualifier=o,value=
45.         qualifier=v,value=
46.         */
47.        filters.clear();
48.
49.        /**
50.         * 查询表 s_behavior 满足如下条件的数据行：
51.         * 1.列限定符 pc:o 值等于 1004 或者 1002
52.         * 2.列数据时间戳等于 1523541994550L
53.         * 3.只返回数据的行键，不返回列数据值
54.         */
55.        SingleColumnValueFilter singleColumnValueFilter2 = new
    SingleColumnValueFilter(Bytes.toBytes(CF_PC), Bytes.toBytes(COLUMN_ORDER),
56.            CompareFilter.CompareOp.EQUAL, Bytes.toBytes("1002"));
57.        filters.add(singleColumnValueFilter);
58.        filters.add(singleColumnValueFilter2);
59.        FilterList filterListAll = new FilterList(FilterList.Operator.MUST_PASS_ALL,
60.            new KeyOnlyFilter(),
61.            timestampsFilter,
62.            new FilterList(FilterList.Operator.MUST_PASS_ONE, filters));
63.        Scan scanWithValue = new Scan();
64.        scanWithValue.setFilter(filterListAll);
65.        resultScanner = table.getScanner(scanWithValue);
66.        printResult(resultScanner);
67.        /**
68.         输出结果：
69.         rowkey=5432100000000000000009223370513312200861
70.         qualifier=v,value=1002
71.         rowkey=5432100000000000000009223370513312200862
```

```
72.            qualifier=o,value=1004
73.            qualifier=v,value=1004
74.        */
75.    }
76.
77. }
```

如果将第 59 行代码改为代码清单 6-23 所示的样子，则 `KeyOnlyFilter` 又会生效。

代码清单 6-23　FilterList 使用 2
```
1.        FilterList filterListAll = new FilterList(FilterList.Operator.MUST_PASS_ALL,
2.            timestampsFilter,
3.            new FilterList(FilterList.Operator.MUST_PASS_ONE, filters),
4.            new KeyOnlyFilter());
```

HBase 提供了很多过滤器，通常情况下已经足够使用，但是我们仍然可以通过继承 HBase 提供的过滤器接口 `Filter` 和抽象类 `FilterBase` 来定义自己需要的过滤器，从使用类型上来说过滤器可以分为以下 3 类。

- 比较过滤器：包括 `RowFilter`、`FamilyFilter` 和 `ValueFilter` 等构造函数需要接受一个比较器，并且继承自 `CompareFilter` 的过滤器。
- 专用过滤器：包括 `SingleColumnValueFilter`、`PageFilter`、`PrefixFilter` 和 `KeyOnlyFilter` 等用来实现某个特定操作的过滤器。
- 附加过滤器：包括 `SkipFilter` 和 `WhileMatchFilter` 这两个包装类型的过滤器。

HBase 过滤器在客户端创建，通过 RPC 请求在客户端与服务端之间传输，在服务端发挥过滤数据的作用，减少了服务端与客户端传输的数据量，并且通过过滤器的组合实现了获取特定数据的目的，是 HBase 数据查询不可或缺的重要一环。

6.5　事务

先回顾下关系型数据库的事务的 4 个属性，通常称之为 ACID。

- 原子性（atomicity）：一个事务是一个不可分割的操作单元，要么一起成功，要么一起失败。
- 一致性（consistency）：事务必须使数据库从一个一致性状态变到另一个一致性状态，这个是由原子性保证的。
- 隔离性（isolation）：如果一个操作完全不会受到其他任何并发的操作或者事务的影响，那么这个操作是隔离的。SQL 标准描述了事务的 4 个隔离级别，即读未提交（Read Uncommitted）、读已提交（Read Committed）、可重复读（Repeatable Read）和串行（Serializable）。有些数据库如 Oracle 还提供了只读（Read Only）这个隔离级别。

- 持久性（Durability）：事务一旦提交，对数据的改变是永久的。

事务的隔离级别描述了通过锁的粒度来控制并发能力，ANSI/ISO SQL 定义的标准隔离级别有 4 种，从高到底依次为串行（Serializable）、可重复读（Repeatable Read）、读已提交（Read Committed）和读未提交（Read Uncommitted），并发能力则相反，4 个级别从低到高，即串行最低，读未提交最高。

关于事务的 4 种隔离级别分别解决的诸如脏读、可重复读、幻象读这几个问题这里不再赘述，如果读者对此不熟悉或者有疑问可以查询数据库相关资料先行学习。

HBase 作为分布式数据库的一员，必然也是支持事务的，但是其支持的事务范围比较小，默认只能做到支持行级事务。一致性与持久性是事务的基本保障，接下来从原子性与隔离性来分析 HBase 对事务的支持。

6.5.1 原子性

HBase 支持行级事务，即 HBase 对一行数据的操作能够保证原子性。

- HBase 对一行的操作（如 Put）能够保证原子性。一个 Put 操作会有一个返回值——成功、失败或者超时。如果请求超时，那么这个 Put 可能成功也可能失败，总而言之就是要么一行的所有列族所有列都 Put 成功，要么所有列族所有列全部失败。
- HBase 不能保证多行的更新操作的原子性，如一个批量操作 Put a、b 和 c 3 行数据，结果可能是 a、b 成功，c 失败。
- CAS（Compare And Set）API（如 checkAndPut）是一个原子操作。

HBase 事务的原子性是通过 WAL 来保证的，数据写入时会先写入 WAL，再写入 MemStore。写入 MemStore 异常可以通过 WAL 来回滚或者重做，因此只需要保证 WAL 的原子性。假设现在需要写入一行数据 R1，该行数据包含 a、b 和 c 这 3 列，那么 WAL 事务单元的格式为：

```
<logseq#-for-entire-txn>:<-1, 3, <Keyvalue-for-edit-a>, <KeyValue-for-edit-b>,
<KeyValue-for-edit-c>>
```

通过将所有的列写入一行 WAL 事务单元，HBase 保证了一行数据操作的原子性。

6.5.2 隔离性

HBase 支持两种事务隔离级别，类 org.apache.hadoop.hbase.client.IsolationLeve 里面定义了这两种隔离级别：

- READ_COMMITTED(1);
- READ_UNCOMMITTED(2)。

下面通过一个例子来看一下这两个隔离级别的不同，如代码清单6-24所示。

代码清单6-24　HBase事务支持

```
1.   package com.mt.hbase.chpt06.clientapi;
2.
3.   import com.mt.hbase.connection.HBaseConnectionFactory;
4.   import org.apache.hadoop.hbase.Cell;
5.   import org.apache.hadoop.hbase.CellUtil;
6.   import org.apache.hadoop.hbase.TableName;
7.   import org.apache.hadoop.hbase.client.*;
8.   import org.apache.hadoop.hbase.util.Bytes;
9.
10.  import java.util.ArrayList;
11.  import java.util.List;
12.
13.  public class HbaseTransactionDemo {
14.
15.      private static final String TABLE         = "s_test";
16.      private static final String COLUMN_FAMILY = "cf";
17.
18.
19.      private void doPut(String valuePrefix) throws Exception {
20.          List<Put> actions = new ArrayList<Put>();
21.          for (int i = 0; i < 1000; i++) {
22.              Put put = new Put(Bytes.toBytes("rowkey" + i));
23.              put.addColumn(Bytes.toBytes(COLUMN_FAMILY), Bytes.toBytes("a"),
24.                  Bytes.toBytes(valuePrefix + i));
25.              actions.add(put);
26.          }
27.          Table table =
         HBaseConnectionFactory.getConnection().getTable(TableName.valueOf(TABLE));
28.          Object[] results = new Object[actions.size()];
29.          table.batch(actions, results);
30.          System.out.println("put done");
31.      }
32.
33.
34.      private void doScan() throws Exception {
35.          Scan scan = new Scan();
36.          scan.setCaching(1);// 表示一个RPC请求只取一条数据
37.          scan.setIsolationLevel(IsolationLevel.READ_UNCOMMITTED);// 设置事务隔离
         级别为读未提交
38.          Table table =
         HBaseConnectionFactory.getConnection().getTable(TableName.valueOf(TABLE));
39.          ResultScanner resultScanner = table.getScanner(scan);
40.          Result result = null;
41.          while ((result = resultScanner.next()) != null) {
42.              if (result.getRow() == null) {
43.                  continue;// keyvalues=NONE
```

```
44.             }
45.             Cell[] cells = result.rawCells();
46.             for (Cell cell : cells) {
47.                 String qualifier = Bytes.toString(CellUtil.cloneQualifier(cell));
48.                 String value = Bytes.toString(CellUtil.cloneValue(cell));
49.                 System.out.println(
50.                     Bytes.toString(result.getRow()) + "[" + qualifier + "=" + value + "]");
51.             }
52.         }
53.     }
54.
55.     public static void main(String[] args) throws Exception {
56.         final HbaseTransactionDemo hbaseTransactionTest = new HbaseTransactionDemo();
57.         hbaseTransactionTest.doPut("t1");
58.
59.         Thread putThread = new Thread(new Runnable() {
60.             @Override public void run() {
61.                 try {
62.                     hbaseTransactionTest.doPut("t2");
63.                 } catch (Exception e) {
64.                     e.printStackTrace();
65.                 }
66.             }
67.         });
68.         putThread.start();
69.
70.         Thread scanThread = new Thread(new Runnable() {
71.             @Override public void run() {
72.                 try {
73.                     hbaseTransactionTest.doScan();
74.                 } catch (Exception e) {
75.                     e.printStackTrace();
76.                 }
77.             }
78.         });
79.         scanThread.start();
80.     }
81.
82. }
83.
```

代码清单 6-24 解释如下。

（1）Put 1000 行数据到 t_test 表的 cf 列族，行键是"行键+<<序号>>"。每行有一个列，列限定符是 a，列值是"t1+<<序号>>"。

（2）新起一个名为 putThread 的线程，该线程所做的事情与第 1 步相同，不同点是 Put 的 a 列的列值是"t2+<<序号>>"。

（3）新起一个名为 `scanThread` 的线程，该线程发起一个 Scan 请求查询 t_test 表，注意如下两行代码：

```
scan.setCaching(1);// 表示一个 RPC 请求只取一条数据
scan.setIsolationLevel(IsolationLevel.READ_UNCOMMITTED);// 设置事务隔离级别为读未提交
```

如果扫描的数据量比较大，可能会有多次 RPC 调用。第一行设置 1 次 RPC 调用只查询一行数据，这个为了让 Scan 请求的跨度更长，这样 `putThread` 事务就可能与 `scanThread` 事务交叉，`putThread` 就可能会修改到 Scan 请求需要读取而未读取到的数据。第二行设置 Scan 的事务隔离级别为读未提交。

代码清单 6-24 输出结果片段如代码清单 6-25 所示。

代码清单 6-25　代码清单 6-24 输出结果 1

```
1.  put done
2.  rowkey0[a=t10]
3.  rowkey1[a=t11]
4.  rowkey10[a=t110]
5.  rowkey100[a=t2100]
6.  rowkey101[a=t2101]
7.  rowkey102[a=t2102]
8.  rowkey103[a=t2103]
9.  rowkey104[a=t2104]
10. rowkey105[a=t2105]
11. rowkey106[a=t2106]
12. rowkey107[a=t2107]
13. rowkey108[a=t2108]
14. rowkey109[a=t2109]
15. rowkey11[a=t211]
16. rowkey110[a=t2110]
17. put done
18. rowkey111[a=t2111]
19. ...
20.
```

注意，输出结果的第 4 和 5 行，可以看到 `rowkey100` 这条记录，列 a 的值前缀是"t2"，说明 Scan 请求的结果已经被 `putThread` 的修改影响到了，也就是所谓的"脏读"。

接下来将代码清单 6-24 的第 37 行代码改为如下内容，将 Scan 的事务隔离级别改为读已提交：

```
scan.setIsolationLevel(IsolationLevel.READ_COMMITTED);
```

再次运行代码清单 6-24，输出结果如代码清单 6-26 所示。

代码清单 6-26　代码清单 6-24 输出结果 2

```
1.  put done
2.  rowkey0[a=t10]
```

```
3.    rowkey1[a=t11]
4.    rowkey10[a=t110]
5.    rowkey100[a=t1100]
6.    rowkey101[a=t1101]
7.    rowkey102[a=t1102]
8.    rowkey103[a=t1103]
9.    rowkey104[a=t1104]
10.   rowkey105[a=t1105]
11.   rowkey106[a=t1106]
12.   rowkey107[a=t1107]
13.   put done
14.   rowkey108[a=t1108]
15.   ...
16.   rowkey152[a=t1152]
17.   rowkey153[a=t1153]
18.   rowkey154[a=t1154]
19.   rowkey155[a=t1155]
20.   rowkey156[a=t1156]
21.   ...
22.   rowkey997[a=t1997]
23.   rowkey998[a=t1998]
24.   rowkey999[a=t1999]
25.
```

从上面的输出结果可以看到，scanThread 的结果并未受 putThread 影响（列限定符 a 的值的前缀一直是"t1"），而且在 putThread 执行完成之后，scanThread 查询到的结果仍然是 scanThread 开始时的值。

HBase 读已提交的隔离级别也是由一个称之为"多版本读一致"的功能支持的，这在关系型数据库（如 Oracle）中很常见，HBase 的实现方式如下。

（1）当一个写事务（如一组 Put 或者 Delete）开始的时候，首先会获取一个事务编号，这里称之为 WriteNumber，WriteNumber 是一个自增数字，这个自增数字在分区级别自增并且唯一，写入的每一个键值单元值都会有一个 WriteNumber。

（2）当一个读事务（如 Scan 或者 Get）开始的时候，如果事务隔离级别为 ReadCommited，那么会获取到上一次提交事务的事务编号，这里称之为 ReadPoint。

（3）扫描结果过滤掉所有 WriteNumber 大于 ReadPoint 的键值。

当然，实际情况会复杂很多。例如，怎样去构建数据扫描器（StoreFileScanner 和 MemStoreScanner 等）、Scan 怎样根据过滤器过滤数据等。从上述流程可以看出，HBase 可以在没有加锁的情况下通过多版本的控制，保证在读已提交事务隔离级别下数据的一致性，即数据在读事务开始时刻（扫描器构建的时刻）就已经决定了。

第 7 章

架构实现

对于一个数据库,如果不了解其数据存储结构、数据查询流程,那就只会停留在简单使用上,"知其然而不知其所以然"。本章介绍 HBase 的存储结构与数据写入和查询流程,期望读者读完本章后能够做到"知其然而知其所以然"。

7.1 存储

不同的排序、搜索算法在性能上差别可以非常大,从 $O(n^2)$ 到 $O(\log_2 n)$ 甚至 $O(1)$。同样,不同的存储模型和索引结构对数据库的读写性能影响很大。本节通过 B+树与 LSM 存储模型来介绍 HBase 如何在提供高写性能的同时也能够保证读性能。

7.1.1 B+树

二叉查找树(binary search tree)是数据结构与算法里比较经典的查找数据结构,其查找时间复杂度 $O(\log_2 n)$ 与树的深度相关。二叉查找树的特点是每个非叶子节点都只包含两个子节点,当数据量非常大时,树的深度很深,二叉查找树搜索时需要访问的节点就多,这些节点数据通常都存储在磁盘,那么每一次节点访问就相当于进行了一次磁盘 IO 操作。树的深度越深,需要访问磁盘的次数越多,搜索效率就越慢。

B 树(balanced tree)平衡树在二叉树的基础上优化而来,主要目的是降低树的深度,减少磁盘 IO。B 树通过改二叉为多叉,在每个节点上存储更多的指针信息,降低磁盘 IO 数目。B 树的所有叶子节点都具有相同的深度,因此数据查找时间比较平衡。图 7-1 展示了一个深度为 2 的 B 树(这里定义根节点的深度为 0)。

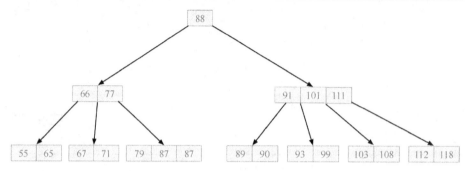

图 7-1　B 树

B+树是 B 树的一种变体，其与 B 树的区别如下。
- 叶子节点包含了全量的索引数据，一般会存储一个指向数据的指针，如 Oracle 的 RowID。
- 叶子节点按照索引的顺序从小到大连接起来。
- 所有的非叶子节点包含了一部分的索引数据，节点的索引数据为其子节点中最大或者最小的索引数据。

图 7-2 表示了一个深度为 2 的 B+树。

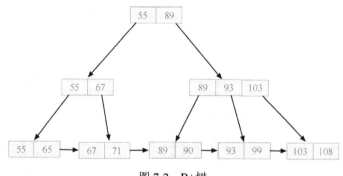

图 7-2　B+树

相对来说，B+树比 B 树更适合做数据库系统索引，尤其是对于 HBase 这类常用区间扫描的数据库，因为 B+树只需要遍历所有叶子节点即可实现区间扫描而无须回溯到父节点甚至根节点。B+树最大的性能问题是插入，随着越来越多数据的插入，叶子节点会慢慢分裂，逻辑上连续的叶子节点可能会存储到磁盘的不同块，做区间扫描时可能产生大量随机读 IO，同时数据写入时维护树的分裂、合并也会产生大量随机写 IO。

7.1.2　LSM 树

为了解决 B+树随机写 IO 过多的问题，HBase 引入了 LSM 树（log structured

merge-trees），LSM 的核心思想是将一棵大树拆分为多棵小树，HBase 数据的写入都会先写 MemStore，在内存中构建一棵有序的小树，当 MemStore 达到一定条件时即会刷新输出写入到磁盘为一个 StoreFile，此时的数据已经有序并且因为是顺序写入磁盘，所以写入速度很快。这里的弊端是随着数据量的增大，StoreFile（即小树）会越来越多，数据查询的时候需要扫描所有的文件，显然文件越多，扫描效率越低。

LSM 中 M 代表的 merge（合并）就是为了解决 StoreFile 越来越多而造成读性能下降的问题。当 MemStore 刷新后 StoreFile 达到配置的数量或者距离上次压缩时间满足配置的间隔时，HBase 即会自动触发压缩（分为 `MINOR` 和 `MAJOR` 两种压缩类型，另一种说法叫作大小合并）。图 7-3 展示了树 T-M 从内存 MemStore 刷新到磁盘为 StoreFile，之后多个 StoreFile 存储的小树合并为一棵大树的过程。

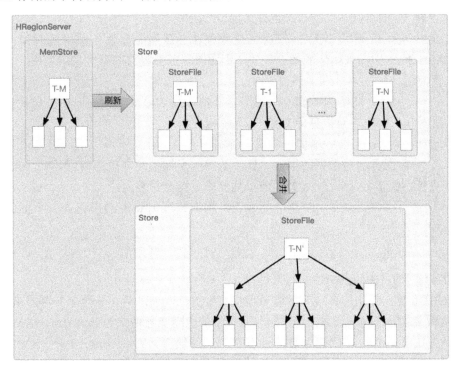

图 7-3　MemStore 刷新与 StoreFile 合并

LSM 的核心思想是通过牺牲一定的读性能来换取写能力的最大化。为了避免读性能下降而成为瓶颈，HBase 也提供了一些其他策略，例如使用布隆过滤器（详见第 9 章）等来过滤数据查询需要读取的文件数量以提高读性能。

LSM 为了提高写性能，数据先写入内存。如果服务器宕机或者断电等，就会导致内存数据丢失，为了解决该问题，HBase 引入了 WAL。

7.1.3 WAL

WAL（write ahead log）即预写入日志。由于 LSM 为了提高写性能，数据先写入内存，如果服务器宕机或者断电等会导致内存数据丢失，因此类似于 Oracle 事务提交之前需要写重做（redo）日志，HBase 数据写入内存之前也需要写 WAL 以便异常恢复。

WAL 有以下几种持久化类型（见 `org.apache.hadoop.hbase.client.Durability`）。

- `SYNC_WAL`：将数据的修改以同步的方式写入 WAL，并且刷新到文件系统，如刷新为 HDFS 的 HLog。
- `USE_DEFAULT`：使用默认的持久化策略，HBase 全局默认的持久化策略为 `SYNC_WAL`。
- `SKIP_WAL`：不写 WAL，该持久化类型在分区服务器发生故障的时候可能导致 MemStore 中尚未刷新的数据丢失，因此除非业务能够容忍数据的丢失，否则不建议使用。
- `ASYNC_WAL`：将数据的修改以异步的方式写入 WAL，与 `SYNC_WAL` 相比性能会有所提升，客户端无须等待 WAL 写入完成后返回。
- `FSYNC_WAL`：当前版本仍未实现，等同于 `SYNC_WAL`。

HBase 分区服务器对数据的插入、删除等每一个键值的修改都会封装到 `WALEdit`，HBase 仅支持行级事务，因此一行数据的多列的修改均会封装到同一个 `WALEdit`，`WALEdit` 的格式示例如下（最开始的 `-1` 是一个标识，用来兼容旧版本的只包含一个键值修改的 WAL）：

```
<-1, 3, <KV-for-edit-c1>, <KV-for-edit-c2>, <KV-for-edit-c3>>
```

图 7-4 演示了 HBase WAL 的写入过程。

当 HBase 客户端向服务端分区服务器提交数据修改请求后，分区服务器的 RPC 处理线程池负责接收处理这些请求，客户端与服务端的数据交互格式为 Protobuf。在 WAL 的写入过程中使用了一个叫作 LMAX Disruptor 的高性能缓存队列 `RingBuffer`，`RPCHandler` 作为生产者向 `RingBuffer` 按序列（sequence）添加数据变更封装成的 `WALEdit`，而 `FSHLog` 的内部类 `RingBufferEventHandler` 则作为消费者从队列读取数据然后写入文件系统（一般存储到 HDFS 系统），这个生产消费模型是一个多生产者单消费者模型，单消费者按 `RingBuffer` 序列顺序刷新数据到文件系统，保证 WAL 并发写入只有一个线程在真正写入文件，做到分区服务器全局唯一，WAL 的实现类为 `FSHLog.java`。以下步骤描述了整个 WAL 从写入到刷新数据到文件系统，再到通知写入成功的过程。

（1）每个分区都有一个 `wal` 的实例变量，实现类为 `FSHLog.java`。

（2）如果客户端设置的 WAL 持久化策略不为 `Durability.SKIP_WAL`，则 `RPCHandler`

在 HRegion 类将变更数据写入 MemStore 之前会调用 wal.append(..walKey,walEdit,…) 方法，在 wal 的实现类 FSHLog 中该方法将 walEdit 封装为 FSWALEntry，再将 FSWALEntry 装载到一个载体 RingBufferTruck，最后调用 FSHLog 的实例变量 disruptor 将载体 RingBufferTruck 发布到队列，之后不阻塞并继续执行将变更写入 MemStore 等操作。

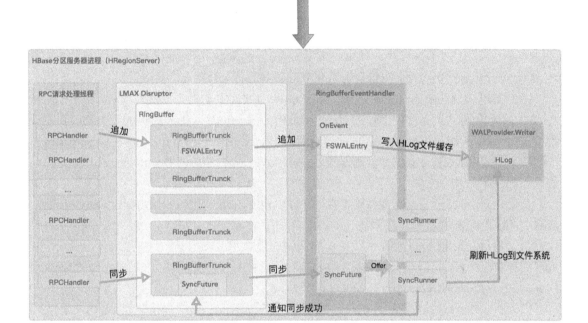

图 7-4　WAL 写入过程

（3）当 RPCHandler 在 HRegion 类将变更数据写入 MemStore 之后，会调用 wal.sync(txid) 方法，将一个 SyncFuture 装载到 RingBufferTruck 对象后调用 FSHLog 的实例变量 disruptor 将载体 RingBufferTruck 发布到队列，这个 SyncFuture 带有一个 RingBuffer 的序列，此时 RPCHandler 线程会阻塞等待 wal 刷新到文件系统，默认等待超时时间为 5 分钟。

（4）当生产者将 RingBufferTruck 发布到 disruptor 的队列后，消费者（FSHLog 的内部类 RingBufferEventHandler）的 onEvent 方法会触发消费队列中的对象 RingBufferTruck。

如果 RingBufferTruck 载体装载的对象为 FSWALEntry，则 RingBufferEventHandler 会将载体携带的 WALEdit 写入缓冲区，处理步骤如下。

（1）从传入的 FSWALEntry.WALKey 获取到 HRegion 的多版本一致性控制器

MultiVersionConcurrencyControl 实例 mvcc，然后调用 mvcc 的 begin() 方法新起一个事务，并得到一个 writeEntry，将 writeEntry 加入 mvcc 的 writeQueue 队列，writeEntry 包含一个递增唯一的 writeNumber 作为事务 ID（假设为 mvccNumX），之后将修改的数据键值对全部写上这个事务 ID mvccNumX（用来做多版本读控制），最后将 writeEntry 赋值给 append 方法传入的 FSWALEntry.WALKey 用来在请求最后完成 mvcc 事务。

（2）调用配置的 WALObserver 协处理器的 preWALWrite 方法，可以用来实现用户自定义的逻辑。调用 WALActionsListener 列表（有 MetricsWAL、Replication 和 LogRoller 这 3 个需要执行）的 visitLogEntryBeforeWrite 方法，实际上只有 Replication 实现了该方法，用来获取 WALKey 的复制 SCOPE（在建表 DDL 指定的 REPLICATION_SCOPE），即判断是否需要复制。

（3）调用 writer.append 方法，writer 的默认实现类为 ProtobufLogWriter.java。如果设置了 WAL 的压缩，该方法会对 WALEntry 写入压缩相关的上下文，再转调 WALCellCoder.EnsureKvEncoder，将数据写入 HdfsDataOutputStream 的缓存区。

（4）调用 WALObserver 协处理器的 postWALWrite 方法，调用 WALActionsListener 的 postAppend 方法。

如果 RingBufferTruck 载体装载的对象为 SyncFuture，则 RingBufferEventHandler 会将缓冲区的数据刷新到文件系统，处理步骤如下：

（1）将 SyncFuture 放入 FSHLog 的实例变量数组 syncFutures，从 SyncRunner 处理线程数组中轮询取一个 SyncRunner，将 SyncFuture 放入该 SyncRunner 的处理队列 BlockingQueue syncFutures。

（2）SyncRunner 处理线程的 run 方法阻塞在从 syncFutures 队列 take 方法。当第 1 步的 SyncFuture 对象被 offer 到 syncFutures 队列后，run 方法被唤醒，拿到该 SyncFuture 对象，首先判断需要处理的 RingBuffer 序列是否小于 FSHLog 实例变量 highestSyncedSequence（用来保存最大已经刷新到文件系统的 RingBuffer 的序列），如果小于，则说明该 WALEdit 变更已经刷新到文件系统，因而将 SyncFuture 置为完成状态，并唤醒前面阻塞的 RPCHandler 线程，然后跳出流程，继续阻塞在从 syncFutures 队列 take 方法。

（3）如果 RingBuffer 序列大于 FSHLog 实例变量 highestSyncedSequence，那么调用 ProtobufLogWriter 的 sync 方法将缓存中的变更刷新到文件系统。注意，因为这里会刷新缓存中的所有变更，所以一些小于当前序列的变更也会被刷新到文件系统（也就是做到了缓冲区的批量刷新），最后将 highestSyncedSequence 更新为当前序列。

（4）将当前序列对应的 SyncFuture 置为完成状态，将对应的序列小于当前序列的 SyncFuture 状态置为完成状态（其实这一步可有可无，第 2 步同样会做该操作，但是这里做了可能对性能会有稍许提升），最后唤醒等待的 RPCHandler 线程。

（5）调用 `WALActionsListener` 的 `postSync` 回调方法。

至此 WAL 已经写入并刷新到文件系统，可以看到 WAL 写入文件系统的时候会阻塞客户端的请求，磁盘的写入效率对 HBase 整体性能影响很大，因此一般都建议对基于 HDFS 存储的 HBase 集群配置多块磁盘以提高性能。

7.2 数据写入读取

HBase 数据的写入与读取需要客户端先通过读取 HMaster 节点上的元数据定位到本次插入或者读取数据所在分区由哪个分区服务器负责，之后 HBase 客户端直接与定位到的分区服务器通信。

7.2.1 定位分区服务器

分区是 HBase 负载均衡的最小单元，均衡的分布在 HBase 集群的每台分区服务器。HBase 客户端进行数据查询、修改、删除等操作时都需要先定位操作数据所在的分区，以及分区由哪个分区服务器负责。HBase 依赖 ZooKeeper 来实现分区服务器的定位，如图 7-5 所示，最多经过 3 步，HBase 客户端即可定位到负责数据操作的分区服务器。

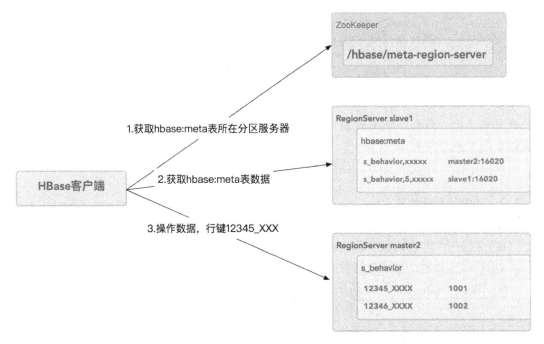

图 7-5 定位分区

（1）HBase 客户端根据连接配置的 ZooKeeper 地址获取到 znode `/hbase/meta-region-server` 的值，即 `hbase:meta` 分区所在的分区服务器地址。注意，`hbase:meta` 有且仅有一个分区，获取到地址后客户端会缓存该地址。

（2）HBase 客户端根据第 1 步获取到的分区服务器地址获取 `hbase:meta` 表数据并缓存，`hbase:meta` 表数据列出了每个分区所在的分区服务器以及开始和结束行键，如表 7-1 所示，同时 HBase 客户端会缓存这些信息。

（3）HBase 客户端将操作的数据根据行键所在的分区服务器分组，分别向对应的分区服务器发起 RPC 请求进行数据操作。由于前面已经缓存了分区的开始结束行键以及所在分区服务器地址，因此之后的数据操作请求只需与对应的分区服务器交互，除非 HBase 客户端捕捉到 `IOException`（如 `WrongRegionException` 和 `RegionMovedException`），此时 HBase 客户端会清除缓存，重新拉取分区位置信息。

表 7-1　hbase:meta 表数据示例

行键	info:regioninfo	info:server	...
s_behavior,,151937210 0459. ef6c8d9ffb26a26e0dd 534845fef21d3.	{ENCODED => ef6c8d9ffb26a26e0dd534845fef21d3, NAME => 's_behavior,,1519372100459. ef6c8d9ffb26a26e0dd534845fef21d3.', STARTKEY => '', ENDKEY => '5'}	master2:16020	...
s_behavior,5,1519372 100459. fe1d8a4df4f1008b28b 313617554245e.	{ENCODED => fe1d8a4df4f1008b28b313617554245e, NAME => 's_behavior,5,1519372100459. fe1d8a4df4f1008b28b313617554245e.', STARTKEY => '5', ENDKEY => ''}	slave1:16020	...

> **注意**　HBase 客户端会将 `hbase:meta` 表数据缓存在本地，因此大部分情况下前两步只有在客户端第一次做数据操作请求的时候发生，因而对 ZooKeeper 集群的压力很小。

7.2.2　数据修改流程

HBase 提供的客户端数据修改操作类包括 `Append`、`Delete`、`Put` 和 `Increment`，这几个数据修改操作类均继承自抽象类 `Mutation`，如图 7-6 所示。

当 HBase 分区服务器启动的时候会初始化一个 `RSRpcService` 类型的实例变量 `rpcServices`，`rpcServices` 初始化的时候会初始化一个基于 Java NIO 的 `Listener` 用来处理客户端的 RPC 请求，HBase 客户端服务端数据交互格式基于 Protobuf。由于 `Put` 与 `Delete` 请求相对比较复杂，而且更常用、更具有代表性，因此接下来以 `Put` 与 `Delete`

请求来分析 HBase 在服务端对数据修改的处理流程，如图 7-7 至图 7-9 所示。

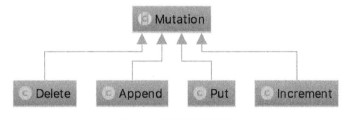

图 7-6　数据修改类图

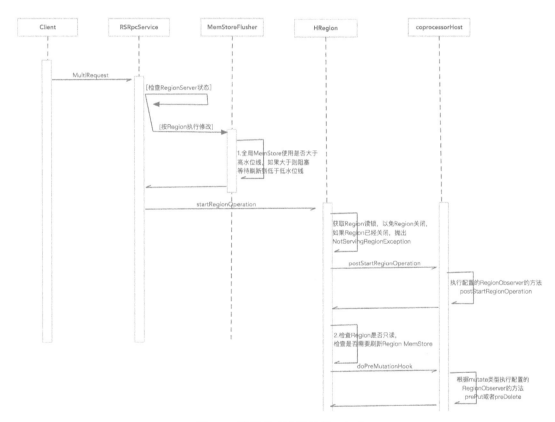

图 7-7　数据修改流程的第一部分

如果全局 MemStore 占用内存大小大于高水位线 A（A=JVM 堆大小×配置项 `hbase.regionserver.global.memstore.size` 或者 `hbase.regionserver.global.memstore.upperLimit` 的值），则触发全局 MemStore 刷新并且阻塞请求，直到全局 MemStore 占用内存大小小于 A；如果全局 MemStore 占用内存大小小于 A 但是大于低水位线 B（B=A×配置项 `hbase.regionserver.global.memstore.size.lower.limit` 的

值),也会触发全局 MemStore 刷新,但是不会阻塞。注意,如果是对元数据表 `hbase:meta` 的修改,则不检查该全局内存占用。

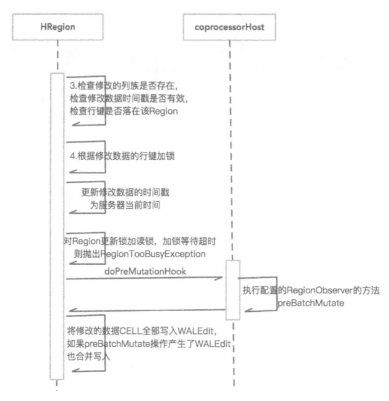

图 7-8 数据修改流程的第二部分

如果分区的 MemStore 占用内存大小大于 B(B=配置项 `hbase.hregion.memstore.flush.size` 的值×配置项 `hbase.hregion.memstore.block.multiplier` 的值),则触发分区的 MemStore 刷新,同时抛出 `RegionTooBusyException` 异常并返回请求失败。

对于 `Put` 操作需要检查操作数据单元格 `Cell` 客户端时间戳是否大于"服务端时间戳+timestampSlop",timestampSlop 为配置项 `hbase.hregion.keyvalue.timestamp.slop.millisecs` 的值,单位毫秒(ms)。如果大于,则抛出 `FailedSanityCheckException` 异常并返回请求失败。

对于多行数据修改的操作,第一行修改数据加锁会等待一个加锁超时时间(配置项 `hbase.rowlock.wait.duration` 的值,默认 30000 ms),剩下的行加锁不管成功与否会立即返回。如果任意一行加锁失败,则跳出,不对下面的行继续加锁,而是接着只处理加锁成功的行,这就使得客户端要注意,对批量的 Put、Delete 操作需要检查返回结果的执行状态,因为有些操作可能并未执行。

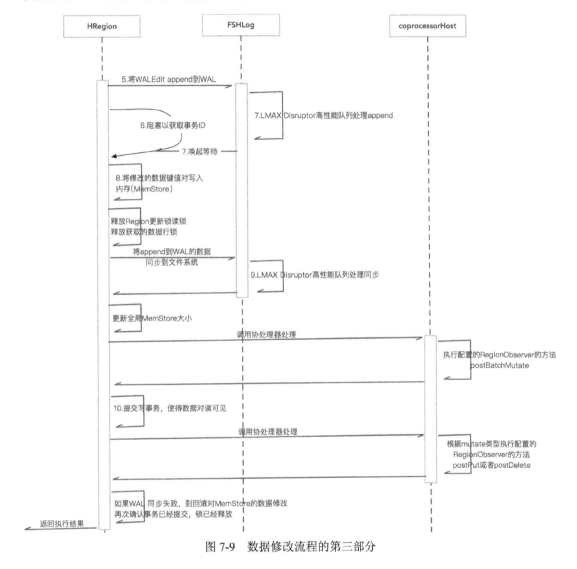

图 7-9 数据修改流程的第三部分

将修改的数据封装成 WALEdit 写入预写入日志 WAL,详情可见 7.1.3 节。

调用 walKey.getWriteEntry() 方法,阻塞在 CountDownLatch 实例 seqNumAssignedLatch 的 wait 方法,目的是为了获得一个全局唯一的事务 ID 作为多版本控制的 mvccNum,用来支持 HBase 的事务隔离,WAL 的 append 动作完成时会设置该 WriteEntry 并唤醒阻塞。

如果客户端设置的 WAL 持久化策略不为 Durability.SKIP_WAL,则调用的两个方

法 `wal.append(..walKey,walEdit,…)` 和 `wal.sync(txid)` 会通过 LMAX Disruptor 的生产消费者队列模型最终调用 `FSHLog` 的内部类 `RingBufferEventHandler` 的方法 `onEvent()`，对于 `append` 事件该方法会从传入的 `walKey` 获取到 `HRegion` 的多版本一致性控制器 `MultiVersionConcurrencyControl` 实例 `mvcc`，然后调用 `mvcc` 的 `begin()` 方法新起一个事务得到一个 `writeEntry A`，将 `writeEntry A` 加入 `mvcc` 的 `writeQueue` 队列，`writeEntry A` 包含一个递增唯一的 `writeNumber` 作为事务 ID（假设为 `mvccNumX`），之后将修改的数据键值对全部写上这个事务 ID `mvccNumX`，最后将 `writeEntry A` 赋值给 `append` 方法传入的 `walKey` 对象，同时唤醒图 7-9 中所示的第 6 步的等待线程。

写入 MemStore 的数据键值对也全部写上了事务 ID，此时这些数据对查询请求仍然不可见，因为 WALEdit 此时并未同步真正刷新到文件系统（如 HDFS）。

接下来写入线会将 WALEdit 真正落地到文件系统。

调用 HRegion 多版本一致性控制器 `MultiVersionConcurrencyControl` 实例 `mvcc` 的 `mvcc.completeAndWait(writeEntry)` 方法，该方法先将 `witeEntry A` 标记为完成，然后遍历 `writeQueue` 队列，检查已经完成的 `writeEntry`，并得到已经完成的 `writeEntry` 的最大事务 ID，假设此时 `writeQueue` 队列只有当前线程添加的 `writeEntry`（即无并发），那么该事务 ID 就是 `mvccNumX`，然后将该分区的全局读取点更新为 `mvccNumX`，并通知其他等待全局读取点更新阻塞的线程。如果此时 `writeQueue` 队列有其他线程添加的未完成的 `writeEntry`，并且该 `writeEntry` 对应的事务 `ID<mvccNumX`，那么当前线程会阻塞等待全局读取点更新，一直等到全局读取点大于等于该 `mvccNumX`（想想这是为什么？答案是为了保证事务的顺序提交）。阻塞的同时会有如代码清单 7-1 所示日志输出，一般是由于磁盘瓶颈导致 WAL 同步刷新到文件系统速度过慢所致。

代码清单 7-1　RPC 请求阻塞待全局事务读取点向上更新

```
2018-02-11 14:01:05,915 WARN  [B.defaultRpcServer.handler=10,queue=10,
port=16020] regionserver.MultiVersionConcurrencyControl: STUCK:
MultiVersionConcurrencyControl{readPoint=45815693, writePoint=45815702}
2018-02-11 14:01:06,039 WARN  [B.defaultRpcServer.handler=73,queue=13,
port=16020] regionserver.MultiVersionConcurrencyControl: STUCK:
MultiVersionConcurrencyControl{readPoint=45815693, writePoint=45815708}
2018-02-11 14:01:06,466 WARN  [B.defaultRpcServer.handler=120,
queue=0,
port=16020] regionserver.MultiVersionConcurrencyControl: STUCK:
MultiVersionConcurrencyControl{readPoint=46352579, writePoint=46352586}
```

整个数据修改流程相对复杂，但是对于深入理解 HBase 的架构设计很有帮助，因此如果读者觉得现在读起来比较晦涩或者枯燥，可以在读完本书后回过头来看看，该建议同样适用于下面的数据查询流程。

7.2.3 数据查询流程

HBase 提供的客户端数据查询操作类包括 `Scan`、`Get`，`Get` 请求会被封装成 `Scan`，然后当成一个特殊的 `Scan` 类型处理，接下来以 `Scan` 为例介绍 HBase 的数据查询流程，图 7-10 和图 7-11 描述了 `Scan` 的流程图。

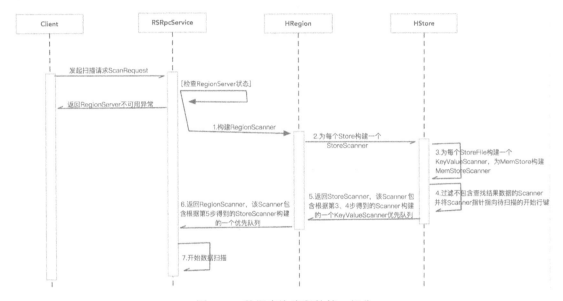

图 7-10 数据查询流程的第一部分

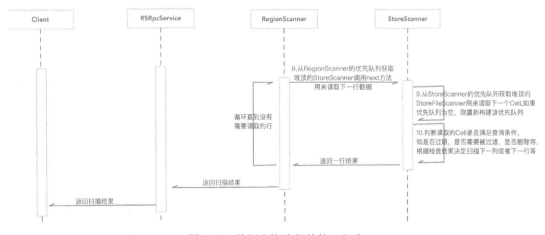

图 7-11 数据查询流程的第二部分

在为 `Scan` 请求构建 `RegionScanner` 前后都会触发协处理器的 `preScannerOpen` 和

`postScannerOpen` 回调方法，默认情况下会构建一个 `RegionScannerImpl` 实例。如果 `Scan` 为倒序扫描，则会构建一个 `ReversedRegionScannerImpl` 实例。

`RegionScanner` 的构造方法会为每个 `Store` 构建一个 `StoreScanner`（因为 HBase 每个列族对应一个 `Store`，所以 HBase 表有几个列族就会构建几个 `StoreScanner`，`StoreScanner` 继承自 `KeyValuerScanner`）。

`StoreScanner` 的构造方法会为每个 `StoreFile` 构建一个 `StoreFileScanner`，为每个 `MemoStore` 构建一个 `MemStoreScanner`（`StoreFileScanner` 和 `MemStoreScanner` 统一继承自 `KeyValuerScanner`）。

过滤第 3 步中构建的 `Scanner`，过滤方式包括：数据的时间范围是否满足 `Scan` 条件、数据的行键是否与 `Scan` 开始与结束行键区间有重叠、是否能够通过布隆过滤器过滤，这一步的作用是用来过滤掉一些不需要扫描的文件或者 `MemStore` 以提高读取效率。

一个 `StoreScanner` 包含一个或多个 `StoreFileScanner`、一个或者零个 `MemStoreScanner`，这些 `Scanner` 存储在一个优先队列 `KeyValueHeap` 中，按 `KeyValue`（继承自 `Cell`，代表一列数据）从小到大排列。

一个 `RegionScanner` 包含一个或多个 `StoreScanner`，同样使用优先队列 `KeyValueHeap` 存储。

数据扫描先从 `RegionScanner` 的 `KeyValueHeap` 出栈一个 `StoreScanner`，然后从 `StoreScanner` 的 `KeyValueHeap` 出栈一个 `StoreFileScanner`(或者 `MemStoreScanner`)，再使用这个 `Scanner` 读取数据，之前 `StoreScanner` 初始化的时候有一个搜寻（seek）过程，会将其对应的 `StoreFileScanner` 都定位到需要扫描的行键位置。

读取一行数据的过程会依次读取这行包含的所有 `KeyValue`，然后做行键过滤、`KeyValue` 过滤等，期间如果发现这行数据不是所需要的，则可以跳出读取下一行。

`StoreScanner` 每次都会从 `KeyValueHeap` 出栈一个 `Scanner` 用来真正读取数据，如果 `KeyValueHeap` 队列全部都出栈完成了，又会以同样的方式构建一个 `KeyValueHeap`。

`StoreScanner` 委托 `StoreFileScanner` 读取 `KeyValue` 后会有一系列的过滤动作，例如在 `StoreFileScanner` 读取 `KeyValue` 的时候会跳过数据事务版本号比当前 `Scan` 新的数据、检查当前 `KeyValue` 是否是 `Scan` 需要返回的列、检查当前 `KeyValue` 是否已经过期等。

这里并没有详细介绍过滤器部分，读者可以在 6.4 节了解到过滤器的原理与作用阶段。

第 8 章

协处理器

　　HBase 作为列式数据库最让开发人员头疼的地方是不支持二级索引，开发人员需要"竭尽所能"地将需要查询的字段放入行键。虽然 HBase 提供了对过滤器的支持来实现关系型数据库 SELECT 语句的查询条件，但是对于统计计数、求和等操作，实现起来仍然非常复杂。

　　很多时候由于这种复杂性导致需要在客户端去做这些计算，因此需要在服务端与客户端之间传输大量的数据，而且有时候由于客户端单机的硬件限制，这些数据是没办法在客户端被很好地处理的，不管是网络 IO 还是客户端内存可能都会是瓶颈，此时 HBase 协处理器就有了用武之地。通过协处理器，我们可以把客户端复杂的计算逻辑移动到分区服务器去处理，即在数据的实际存储位置执行计算，这就是常说的移动"计算"比移动"数据"容易，数据在服务端处理完之后，再将处理结果返回给客户端，这样就避免了大量的网络 IO，效率大大提升。

　　因为协处理器是运行在服务端的，一旦代码有 bug 或者性能问题，对 HBase 集群性能甚至数据完整性等都有可能造成不可恢复的影响，所以建议高级开发人员使用。

　　协处理器可以分为两大类：观察者类型协处理器（observer coprocessors）和端点类型协处理器（endpoint coprocessor）。

8.1 观察者类型协处理器

　　观察者类型协处理器（observer coprocessor）类似于关系型数据库里面的触发器（trigger，在数据插入、删除等之前或者之后执行）或者面向切面编程里面的切面 AOP（在方法执行前后执行）。如图 8-1 所示，观察者类型协处理器提供了接口，使得开发者可以在获取扫描器前后、查询数据之前、查询到每条数据之后等织入自己的数据处理逻辑。

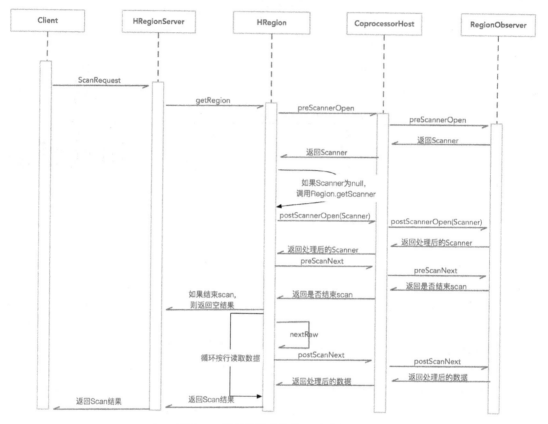

图 8-1　观察者类型协处理器时序图

观察者类型协处理器通常可以用在以下场景。

- 安全：当客户端对数据进行 Get、Put、Scan 等操作时，可以在 RegionObserver 的 preGet、prePut、preScanNext 方法做权限验证（如上时序图，在 preScanNext 方法做权限验证后，如果权限验证失败，则可以直接结束 Scan 并返回空结果），或者对 Get、Scan 的输出结果做安全性过滤，例如对某些安全性较高的字段进行脱敏处理。
- 数据完整性校验：HBase 没有提供类似于关系型数据库的外键等约束，使用 RegionObserver 可以在数据插入之前调用协处理器 prePut 方法，检查外键是否存在。
- 二级索引：HBase 没有提供对二级索引的支持，通过协处理器可以在数据 Put、Delete 时来维护一个自定义的二级索引，华为的开源 HBase 二级索引项目 HIndex 即使用观察者类型协处理器实现。

图 8-2 列出了观察者类型协处理器的类图结构。

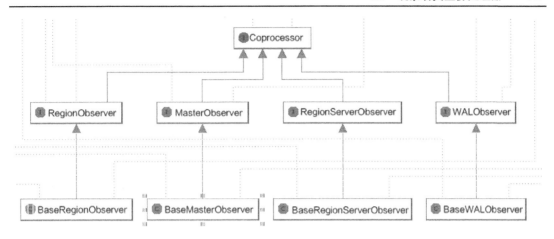

图 8-2 观察者类型协处理器类图

- `MasterObserver`：该接口类定义一些与 HMaster 进程负责指令相关的交互钩子，如建表前后执行的 `preCreateTable` 和 `postCreateTable` 方法，删除表前后执行的 `preDeleteTable` 和 `postDeleteTable` 等 DDL 操作。HBase 提供了一个默认的实现类 `BaseMasterObserver`。
- `RegionServerObserver`：该接口类定义一些与 HRegionServer 进程负责指令相关的交互钩子，如分区服务器停止之前执行的 `preStopRegionServer` 方法，分区合并前后执行的 `preMerge` 和 `postMerge` 方法，以及复制初始化后执行的 `postCreateReplicationEndPoint` 方法，复制数据前后执行的 `preReplicateLogEntries` 和 `postReplicateLogEntries` 方法。HBase 提供了一个默认的实现类 `BaseRegionServerObserver`。
- `RegionObserver`：该接口类定义一些与分区操作相关的交互钩子，如分区对 HMaster 上报开启前后执行的 `preOpen` 和 `postOpen` 方法，MemStore 刷新到文件前后执行的 `preFlush` 和 `postFlush` 方法以及客户端相关数据修改操作前后执行的方法 `preGetOp` 和 `postGetOp` 等。HBase 提供了 `RegionObserver` 的一个抽象类 `BaseRegionObserver` 实现，如果需要自定义 `RegionObserver` 可以继承该抽象类。
- `WALObserver`：该接口类提供了对 WALEdit 写入 WAL 操作的交互钩子，如 WALEdit 写入 WAL 前后执行的 `preWALWrite`、`postWALWrite` 方法。HBase 提供了一个默认的实现类 `BaseWALObserver`。

> **注意** 每个类型的 Observer 都有一个带 "Base" 前缀的类或者抽象类的实现，如 `BaseMasterObserver` 继承自 `MasterObserver`。开发者自定义协处理器的时候一般推荐继承这些 Base 抽象类，否则如果忘记实现某些方法，可能会造成服务器端未知的异常。

数据收集后数据安全以及数据的使用权限控制同样是一个重要的话题。假设用户行为系统需要对数据的查询做一个敏感词过滤，定义字符串"阿里巴巴"是一个敏感词，凡是出现"阿里巴巴"的地方，都应该被替换为"**"，这里可以在服务端通过使用 RegionObserver 协处理器对客户端的数据操作请求做一个处理，代码清单 8-1 实现了该敏感词过滤协处理器，该程序主要实现了 postGetOp 和 postScannerNext 这两个方法，作用是对 Get 和 Scan 出来的结果的所有单元格做一个遍历，如果单元格的值包含敏感词"阿里巴巴"，那么就用"**"替换。

代码清单 8-1 敏感词过滤协处理器

```
1.   package com.mt.hbase.chpt08.coprocessor;
2.
3.   import com.google.common.collect.ImmutableList;
4.   import org.apache.commons.collections.CollectionUtils;
5.   import org.apache.hadoop.fs.FileSystem;
6.   import org.apache.hadoop.fs.Path;
7.   import org.apache.hadoop.hbase.*;
8.   import org.apache.hadoop.hbase.client.*;
9.   import org.apache.hadoop.hbase.coprocessor.ObserverContext;
10.  import org.apache.hadoop.hbase.coprocessor.RegionCoprocessorEnvironment;
11.  import org.apache.hadoop.hbase.coprocessor.RegionObserver;
12.  import org.apache.hadoop.hbase.filter.ByteArrayComparable;
13.  import org.apache.hadoop.hbase.filter.CompareFilter;
14.  import org.apache.hadoop.hbase.io.FSDataInputStreamWrapper;
15.  import org.apache.hadoop.hbase.io.Reference;
16.  import org.apache.hadoop.hbase.io.hfile.CacheConfig;
17.  import org.apache.hadoop.hbase.regionserver.*;
18.  import org.apache.hadoop.hbase.regionserver.compactions.CompactionRequest;
19.  import org.apache.hadoop.hbase.regionserver.wal.HLogKey;
20.  import org.apache.hadoop.hbase.regionserver.wal.WALEdit;
21.  import org.apache.hadoop.hbase.util.Bytes;
22.  import org.apache.hadoop.hbase.util.Pair;
23.  import org.apache.hadoop.hbase.wal.WALKey;
24.
25.  import java.io.IOException;
26.  import java.util.Iterator;
27.  import java.util.List;
28.  import java.util.NavigableSet;
29.
30.  /**
31.   *
32.   * 该Observer的作用是实现对结果关键字的脱敏
33.   *
34.   */
35.  public class KeyWordFilterRegionObserver implements RegionObserver {
36.
37.
```

```
38.      // 敏感词
39.      private static final String ALIBABA = "阿里巴巴";
40.      // 敏感词替换字符串
41.      private static final String MASK = "**";
42.
43.
44.      @Override public void postGetOp(
     ObserverContext<RegionCoprocessorEnvironment> observerContext,
45.        Get get, List<Cell> list) throws IOException {
46.          if (CollectionUtils.isNotEmpty(list)) {
47.            for (int i = 0; i < list.size(); i++) {
48.              Cell cell = list.get(i);
49.              KeyValue keyValue = replaceCell(cell);
50.              list.set(i, keyValue);
51.            }
52.          }
53.      }
54.
55.      @Override public boolean postScannerNext(
56.      ObserverContext<RegionCoprocessorEnvironment> observerContext,
57.      InternalScanner internalScanner, List<Result> list, int i, boolean b)
58.      throws IOException {
59.          Iterator<Result> iterator = list.iterator();
60.          while (iterator.hasNext()) {
61.              Result result = iterator.next();
62.              Cell[] cells = result.rawCells();
63.              if (CollectionUtils.isNotEmpty(list)) {
64.                for (int j = 0; j < cells.length; j++) {
65.                  Cell cell = cells[j];
66.                  KeyValue keyValue = replaceCell(cell);
67.                  cells[j] = keyValue;
68.                }
69.              }
70.          }
71.        return true;
72.      }
73.
74.
75.      private KeyValue replaceCell(Cell cell) {
76.          byte[] oldValueByte = CellUtil.cloneValue(cell);
77.          String oldValueByteString = Bytes.toString(oldValueByte);
78.          String newValueByteString = oldValueByteString.replaceAll(ALIBABA, MASK);
79.          byte[] newValueByte = Bytes.toBytes(newValueByteString);
80.          return new KeyValue(CellUtil.cloneRow(cell),
81.                  CellUtil.cloneFamily(cell), CellUtil.cloneQualifier(cell),
82.                  cell.getTimestamp(), newValueByte);
83.      }
84.
85.
```

```java
86.    @Override public void preGetOp(
       ObserverContext<RegionCoprocessorEnvironment> observerContext,
87.    Get get, List<Cell> list) throws IOException {
88.
89.    }
90.
91.    @Override public void preOpen(
       ObserverContext<RegionCoprocessorEnvironment> observerContext)
92.    throws IOException {
93.
94.    }
95.
96.
97.    @Override public void postOpen(
       ObserverContext<RegionCoprocessorEnvironment> observerContext) {
98.
99.    }
100.
101.
102.   @Override public void postLogReplay(
103.   ObserverContext<RegionCoprocessorEnvironment> observerContext) {
104.
105.   }
106.   .
107.
108.   @Override public InternalScanner preFlushScannerOpen(
109.   ObserverContext<RegionCoprocessorEnvironment> observerContext,
       Store store, KeyValueScanner keyValueScanner,
110.   InternalScanner internalScanner) throws IOException {
111.       return internalScanner;
112.   }
113.
114.
115.   @Override public void preFlush(
       ObserverContext<RegionCoprocessorEnvironment> observerContext)
116.   throws IOException {
117.
118.   }
119.
120.
121.   @Override public InternalScanner preFlush(
122.   ObserverContext<RegionCoprocessorEnvironment> observerContext,
123.   Store store,InternalScanner internalScanner) throws IOException {
124.       return internalScanner;
125.   }
126.
127.
```

```
128.    @Override public void postFlush(
        ObserverContext<RegionCoprocessorEnvironment> observerContext)
129.            throws IOException {
130.
131.    }
132.
133.
134.    @Override public void postFlush(
        ObserverContext<RegionCoprocessorEnvironment> observerContext,
135.    Store store, StoreFile storeFile) throws IOException {
136.
137.    }
138.
139.
140.    @Override public void preCompactSelection(
141.    ObserverContext<RegionCoprocessorEnvironment> observerContext,
142.    Store store, List<StoreFile> list, CompactionRequest compactionRequest)
        throws IOException {
143.
144.    }
145.
146.
147.    @Override public void preCompactSelection(
148.    ObserverContext<RegionCoprocessorEnvironment> observerContext,
149.    Store store,List<StoreFile> list) throws IOException {
150.
151.    }
152.
153.
154.    @Override public void postCompactSelection(
155.    ObserverContext<RegionCoprocessorEnvironment> observerContext, Store store,
156.    ImmutableList<StoreFile> immutableList, CompactionRequest
        compactionRequest) {
157.
158.    }
159.
160.
161.    @Override public void postCompactSelection(
162.    ObserverContext<RegionCoprocessorEnvironment> observerContext,
163.    Store store,ImmutableList<StoreFile> immutableList) {
164.
165.    }
166.
167.
168.    @Override public InternalScanner preCompact(
```

```
169.        ObserverContext<RegionCoprocessorEnvironment> observerContext,
170.        Store store,InternalScanner internalScanner, ScanType scanType,
            CompactionRequest compactionRequest)throws IOException {
171.
172.            return internalScanner;
173.        }
174.
175.
176.        @Override public InternalScanner preCompact(
177.        ObserverContext<RegionCoprocessorEnvironment> observerContext, Store store,
178.        InternalScanner internalScanner, ScanType scanType)
179.        throws IOException {return internalScanner;
180.        }
181.
182.
183.        @Override public InternalScanner preCompactScannerOpen(
184.        ObserverContext<RegionCoprocessorEnvironment> observerContext, Store store,
185.        List<? extends KeyValueScanner> list, ScanType scanType, long l,
186.        InternalScanner internalScanner, CompactionRequest compactionRequest)
187.        throws IOException {
188.            return internalScanner;
189.        }
190.
191.
192.        @Override public InternalScanner preCompactScannerOpen(
193.        ObserverContext<RegionCoprocessorEnvironment> observerContext,
194.        Store store, List<? extends KeyValueScanner> list, ScanType scanType,
195.        long l,InternalScanner internalScanner) throws IOException {
196.            return internalScanner;
197.        }
198.
199.
200.        @Override public void postCompact(
            ObserverContext<RegionCoprocessorEnvironment> observerContext,
201.        Store store, StoreFile storeFile, CompactionRequest compactionRequest)
202.        throws IOException {
203.
204.        }
205.
206.
207.        @Override public void postCompact(
            ObserverContext<RegionCoprocessorEnvironment> observerContext,
208.        Store store, StoreFile storeFile) throws IOException {
209.
```

```
210.        }
211.
212.
213.        @Override public void preSplit(
            ObserverContext<RegionCoprocessorEnvironment> observerContext)
214.        throws IOException {
215.
216.        }
217.
218.
219.        @Override public void preSplit(
            ObserverContext<RegionCoprocessorEnvironment> observerContext,
220.        byte[] bytes) throws IOException {
221.
222.        }
223.
224.
225.        @Override public void postSplit(
            ObserverContext<RegionCoprocessorEnvironment> observerContext,
226.        Region region, Region region1) throws IOException {
227.
228.        }
229.
230.
231.        @Override public void preSplitBeforePONR(
232.        ObserverContext<RegionCoprocessorEnvironment> observerContext,
233.        byte[] bytes,List<Mutation> list) throws IOException {
234.
235.        }
236.
237.
238.        @Override public void preSplitAfterPONR(
239.        ObserverContext<RegionCoprocessorEnvironment> observerContext)
            throws IOException {
240.
241.        }
242.
243.
244.        @Override public void preRollBackSplit(
245.        ObserverContext<RegionCoprocessorEnvironment> observerContext)
            throws IOException {
246.
247.        }
248.
249.
250.        @Override public void postRollBackSplit(
```

```
251.     ObserverContext<RegionCoprocessorEnvironment> observerContext)
         throws IOException {
252.
253.     }
254.
255.
256.     @Override public void postCompleteSplit(
257.     ObserverContext<RegionCoprocessorEnvironment> observerContext)
         throws IOException {
258.
259.     }
260.
261.
262.     @Override public void preClose(
         ObserverContext<RegionCoprocessorEnvironment> observerContext,
263.     boolean b) throws IOException {
264.
265.     }
266.
267.
268.     @Override public void postClose(
         ObserverContext<RegionCoprocessorEnvironment> observerContext,
269.     boolean b) {
270.
271.     }
272.
273.
274.     @Override public void preGetClosestRowBefore(
275.     ObserverContext<RegionCoprocessorEnvironment> observerContext,
276.     byte[] bytes, byte[] bytes1, Result result) throws IOException {
277.
278.     }
279.
280.
281.     @Override public void postGetClosestRowBefore(
282.     ObserverContext<RegionCoprocessorEnvironment> observerContext,
283.     byte[] bytes,byte[] bytes1, Result result) throws IOException {
284.
285.     }
286.
287.
288.
289.
290.
291.     @Override public boolean preExists(
292.     ObserverContext<RegionCoprocessorEnvironment> observerContext,
293.     Get get, boolean b)throws IOException {
```

```java
294.         return false;
295.     }
296.
297.
298.     @Override public boolean postExists(
299. ObserverContext<RegionCoprocessorEnvironment> observerContext,
300. Get get, boolean b)throws IOException {
301.         return false;
302.     }
303.
304.
305.     @Override public void prePut(
     ObserverContext<RegionCoprocessorEnvironment> observerContext,
306. Put put, WALEdit walEdit, Durability durability)
307. throws IOException {
308.
309.     }
310.
311.
312.     @Override public void postPut(
313. ObserverContext<RegionCoprocessorEnvironment> observerContext,
314. Put put, WALEdit walEdit, Durability durability)
315. throws IOException {
316.     }
317.
318.
319.     @Override public void preDelete(
     ObserverContext<RegionCoprocessorEnvironment> observerContext, Delete delete,
320. WALEdit walEdit, Durability durability)
321.         throws IOException {
322.
323.     }
324.
325.
326.     @Override public void prePrepareTimeStampForDeleteVersion(
327. ObserverContext<RegionCoprocessorEnvironment> observerContext,
328. Mutation mutation,Cell cell, byte[] bytes, Get get) throws IOException {
329.
330.     }
331.
332.
333.     @Override public void postDelete(
     ObserverContext<RegionCoprocessorEnvironment> observerContext,
334. Delete delete, WALEdit walEdit, Durability durability)
335. throws IOException {
```

```
336.
337.        }
338.
339.
340.        @Override public void preBatchMutate(
341.    ObserverContext<RegionCoprocessorEnvironment> observerContext,
342.    MiniBatchOperationInProgress<Mutation> miniBatchOperationInProgress)
343.    throws IOException {
344.
345.        }
346.
347.
348.        @Override public void postBatchMutate(
349.    ObserverContext<RegionCoprocessorEnvironment> observerContext,
350.    MiniBatchOperationInProgress<Mutation> miniBatchOperationInProgress)
351.    throws IOException {
352.
353.        }
354.
355.
356.        @Override public void postStartRegionOperation(
357.    ObserverContext<RegionCoprocessorEnvironment> observerContext,
358.    Region.Operation operation) throws IOException {
359.
360.        }
361.
362.
363.        @Override public void postCloseRegionOperation(
364.    ObserverContext<RegionCoprocessorEnvironment> observerContext,
365.    Region.Operation operation) throws IOException {
366.
367.        }
368.
369.
370.        @Override public void postBatchMutateIndispensably(
371.    ObserverContext<RegionCoprocessorEnvironment> observerContext,
372.    MiniBatchOperationInProgress<Mutation> miniBatchOperationInProgress,
373.    boolean b)throws IOException {
374.
375.        }
376.
377.
378.        @Override public boolean preCheckAndPut(
        ObserverContext<RegionCoprocessorEnvironment> observerContext,
379.    byte[] bytes, byte[] bytes1, byte[] bytes2,
380.    CompareFilter.CompareOp compareOp, ByteArrayComparable
381.    byteArrayComparable, Put put, boolean b)throws IOException {
382.        return false;
```

```
383.    }
384.
385.
386.    @Override public boolean preCheckAndPutAfterRowLock(
387. ObserverContext<RegionCoprocessorEnvironment> observerContext,
388. byte[] bytes, byte[] bytes1, byte[] bytes2,
389. CompareFilter.CompareOp compareOp,ByteArrayComparable byteArrayComparable,
        Put put, boolean b) throws IOException {
390.        return false;
391.    }
392.
393.
394.    @Override public boolean postCheckAndPut(
395. ObserverContext<RegionCoprocessorEnvironment> observerContext,
396. byte[] bytes, byte[] bytes1, byte[] bytes2, CompareFilter.CompareOp
397. compareOp,ByteArrayComparable byteArrayComparable, Put put,
398. boolean b) throws IOException {
            return false;
399.    }
400.
401.
402.    @Override public boolean preCheckAndDelete(
403. ObserverContext<RegionCoprocessorEnvironment> observerContext,
404. byte[] bytes,byte[] bytes1, byte[] bytes2, CompareFilter.CompareOp
405. compareOp,ByteArrayComparable byteArrayComparable, Delete delete,
406. boolean b) throws IOException {
            return false;
407.    }
408.
409.
410.    @Override public boolean preCheckAndDeleteAfterRowLock(
411. ObserverContext<RegionCoprocessorEnvironment> observerContext,
412. yte[] bytes,byte[] bytes1, byte[] bytes2, CompareFilter.CompareOp
413. compareOp,ByteArrayComparable byteArrayComparable, Delete delete,
414. boolean b) throws IOException {
            return false;
415.    }
416.
417.
418.    @Override public boolean postCheckAndDelete(
419. ObserverContext<RegionCoprocessorEnvironment> observerContext,
420. byte[] bytes,byte[] bytes1, byte[] bytes2, CompareFilter.CompareOp
421. compareOp,ByteArrayComparable byteArrayComparable, Delete delete,
422. boolean b) throws IOException {
            return false;
423.    }
424.
425.
```

```
426.    @Override public long preIncrementColumnValue(
427.    ObserverContext<RegionCoprocessorEnvironment> observerContext,
428.    byte[] bytes, byte[] bytes1, byte[] bytes2, long l, boolean b) throws IOException {
429.        return 0;
430.    }
431.
432.
433.    @Override public long postIncrementColumnValue(
434.    ObserverContext<RegionCoprocessorEnvironment> observerContext,
435.    byte[] bytes, byte[] bytes1, byte[] bytes2, long l, boolean b,
436.    long l1) throws IOException {
            return 0;
437.    }
438.
439.
440.    @Override public Result preAppend(
        ObserverContext<RegionCoprocessorEnvironment> observerContext,
441.    Append append) throws IOException {
442.        return null;
443.    }
444.
445.
446.    @Override public Result preAppendAfterRowLock(
447.    ObserverContext<RegionCoprocessorEnvironment> observerContext,
448.    Append append) throws IOException {
449.        return null;
450.    }
451.
452.
453.    @Override public Result postAppend(
454.    ObserverContext<RegionCoprocessorEnvironment> observerContext,
455.    Append append,Result result) throws IOException {
456.        return result;
457.    }
458.
459.
460.    @Override public Result preIncrement(
461.    ObserverContext<RegionCoprocessorEnvironment> observerContext,
462.    Increment increment) throws IOException {
463.        return null;
464.    }
465.
466.
467.    @Override public Result preIncrementAfterRowLock(
468.    ObserverContext<RegionCoprocessorEnvironment> observerContext,
```

```
469.    Increment increment)throws IOException {
470.        return null;
471.    }
472.
473.
474.    @Override public Result postIncrement(
475. ObserverContext<RegionCoprocessorEnvironment> observerContext,
476.    Increment increment,Result result) throws IOException {
477.        return result;
478.    }
479.
480.
481.    @Override public RegionScanner preScannerOpen(
482. ObserverContext<RegionCoprocessorEnvironment> observerContext,
483.    Scan scan,RegionScanner regionScanner) throws IOException {
484.        return regionScanner;
485.    }
486.
487.
488.    @Override public KeyValueScanner preStoreScannerOpen(
489. ObserverContext<RegionCoprocessorEnvironment> observerContext, Store store,
        Scan scan, NavigableSet<byte[]> navigableSet, KeyValueScanner keyValueScanner)
490.    throws IOException {
491.        return keyValueScanner;
492.    }
493.
494.
495.    @Override public RegionScanner postScannerOpen(
496. ObserverContext<RegionCoprocessorEnvironment> observerContext,
497.    Scan scan,RegionScanner regionScanner) throws IOException {
498.        return regionScanner;
499.    }
500.
501.
502.    @Override public boolean preScannerNext(
503. ObserverContext<RegionCoprocessorEnvironment> observerContext,
504.    InternalScanner internalScanner, List<Result> list, int i,
505.    boolean hasNext)throws IOException {
506.        return hasNext;
507.    }
508.
509.
510.    @Override public boolean postScannerFilterRow(
511. ObserverContext<RegionCoprocessorEnvironment> observerContext,
512.    InternalScanner internalScanner, byte[] bytes, int i, short i1,
```

```
513.    boolean hasNext)throws IOException {
514.        return hasNext;
515.    }
516.
517.
518.    @Override public void preScannerClose(
519. ObserverContext<RegionCoprocessorEnvironment> observerContext,
520. InternalScanner internalScanner) throws IOException {
521.
522.    }
523.
524.
525.    @Override public void postScannerClose(
526. ObserverContext<RegionCoprocessorEnvironment> observerContext,
527. InternalScanner internalScanner) throws IOException {
528.
529.    }
530.
531.
532.    @Override public void preWALRestore(
533. ObserverContext<? extends RegionCoprocessorEnvironment>
534. observerContext,HRegionInfo hRegionInfo, WALKey walKey, WALEdit
    walEdit) throws IOException {
535.
536.    }
537.
538.
539.    @Override public void preWALRestore(
540. ObserverContext<RegionCoprocessorEnvironment> observerContext,
    HRegionInfo hRegionInfo, HLogKey hLogKey, WALEdit walEdit)
541.    throws IOException {
542.
543.    }
544.
545.
546.    @Override public void postWALRestore(
547. ObserverContext<? extends RegionCoprocessorEnvironment>
548. observerContext,HRegionInfo hRegionInfo, WALKey walKey, WALEdit walEdit)
        throws IOException {
549.
550.    }
551.
552.
553.    @Override public void postWALRestore(
```

```
554.    ObserverContext<RegionCoprocessorEnvironment> observerContext,
        HRegionInfohRegionInfo, HLogKey hLogKey, WALEdit walEdit)
555.        throws IOException {
556.
557.    }
558.
559.
560.    @Override public void preBulkLoadHFile(
561.    ObserverContext<RegionCoprocessorEnvironment> observerContext,
562.    List<Pair<byte[], String>> list) throws IOException {
563.
564.    }
565.
566.
567.    @Override public boolean postBulkLoadHFile(
568.    ObserverContext<RegionCoprocessorEnvironment> observerContext,
569.    List<Pair<byte[], String>> list, boolean b) throws IOException {
570.        return false;
571.    }
572.
573.
574.    @Override public StoreFile.Reader preStoreFileReaderOpen(
575.    ObserverContext<RegionCoprocessorEnvironment> observerContext,
        FileSystem fileSystem, Path path, FSDataInputStreamWrapper
576.    fsDataInputStreamWrapper, long l, CacheConfig cacheConfig,
577.    Reference reference, StoreFile.Reader reader)
578.    throws IOException {
579.        return reader;
580.    }
581.
582.
583.    @Override public StoreFile.Reader postStoreFileReaderOpen(
584.    ObserverContext<RegionCoprocessorEnvironment> observerContext,
        FileSystem fileSystem, Path path, FSDataInputStreamWrapper
585.    fsDataInputStreamWrapper, long l, CacheConfig cacheConfig,
586.    Reference reference, StoreFile.Reader reader)
587.    throws IOException {
588.        return reader;
589.    }
590.
591.
592.    @Override public Cell postMutationBeforeWAL(
593.    ObserverContext<RegionCoprocessorEnvironment> observerContext,
594.    MutationType mutationType, Mutation mutation, Cell cell, Cell newCell)
```

```
595.    throws IOException {
596.        return newCell;
597.    }
598.
599.
600.    @Override public DeleteTracker postInstantiateDeleteTracker(
601. ObserverContext<RegionCoprocessorEnvironment> observerContext,
602. DeleteTracker deleteTracker) throws IOException {
603.        return deleteTracker;
604.    }
605.
606.
607.    @Override public void start(CoprocessorEnvironment coprocessorEnvironment)
        throws IOException {
608.
609.    }
610.
611.
612.    @Override public void stop(CoprocessorEnvironment coprocessorEnvironment)
        throws IOException {
613.
614.    }
615. }
```

具体如何将该协处理器部署到分区服务器执行，在 8.3 节会有详细描述。

8.2　端点类型协处理器

端点类型协处理器（Endpoint Coprocessor）类似于关系型数据库里面的存储过程。例如，针对一些复杂的数据操作，在关系型数据库中使用存储过程实现可以使得数据的处理逻辑全部在服务端实现，同时也可以减少在服务端与客户端之间传输的数据量。

端点类型协处理器可以用来做一些统计的工作，例如计算一个表的行数、某一列金额的总数等。相对观察者类型协处理器对客户端的透明，端点类型协处理器需要客户端显式调用，也类似于客户端需要显示调用关系型数据库的存储过程。

HBase 使用 Google Protobuf 协议用作数据的传输，这对端点类型协处理器的开发有些影响。开发中依赖的 Protobuf 需要与 HBase 依赖的 Protobuf 版本兼容，同时应该尽量使用公用的 API。理想的情况是只依赖接口与使用 Protobuf 生成的数据结构，这样当 HBase 需要升级时能够兼容。

假设现在用户行为日志管理系统需要提供一个平台看板，其中一个功能就是显示总成交额，这里可以使用端点类型协处理器，对订单金额列求和，如代码清单 8-2 所示。

代码清单 8-2 求和端点类型协处理器

```java
1.  package com.mt.hbase.chpt08.coprocessor;
2.
3.  import com.google.protobuf.RpcCallback;
4.  import com.google.protobuf.RpcController;
5.  import com.google.protobuf.Service;
6.  import com.mt.hbase.coprocessor.generated.SumDTO;
7.  import org.apache.hadoop.hbase.Cell;
8.  import org.apache.hadoop.hbase.CellUtil;
9.  import org.apache.hadoop.hbase.Coprocessor;
10. import org.apache.hadoop.hbase.CoprocessorEnvironment;
11. import org.apache.hadoop.hbase.client.Scan;
12. import org.apache.hadoop.hbase.coprocessor.CoprocessorException;
13. import org.apache.hadoop.hbase.coprocessor.CoprocessorService;
14. import org.apache.hadoop.hbase.coprocessor.RegionCoprocessorEnvironment;
15. import org.apache.hadoop.hbase.protobuf.ResponseConverter;
16. import org.apache.hadoop.hbase.regionserver.InternalScanner;
17. import org.apache.hadoop.hbase.util.Bytes;
18.
19. import java.io.IOException;
20. import java.util.ArrayList;
21. import java.util.List;
22.
23. public class SumOrderEndpoint extends SumDTO.SumService implements
    Coprocessor, CoprocessorService {
24.
25.     private RegionCoprocessorEnvironment regionCoprocessorEnvironment;
26.
27.     @Override public void getSum(RpcController controller,
28.         SumDTO.SumRequest request,RpcCallback<SumDTO.SumResponse> done) {
29.
30.         Scan scan = new Scan();
31.         scan.addFamily(Bytes.toBytes(request.getFamily()));
32.         scan.addColumn(Bytes.toBytes(request.getFamily()),
        Bytes.toBytes(request.getColumn()));
33.
34.         SumDTO.SumResponse response = null;
35.         InternalScanner scanner = null;
36.         try {
37.             scanner = regionCoprocessorEnvironment.getRegion().getScanner(scan);
38.             List<Cell> results = new ArrayList<Cell>();
39.             long sum = 0L;
40.             while(true){
41.                 boolean hasMore = scanner.next(results);
42.                 for (Cell cell : results) {
43.                     if(cell.getValueLength() > 0){
44.                         String cellValue = Bytes.toString(CellUtil.cloneValue(cell));
45.                         sum = sum + Long.parseLong(cellValue);
```

```
46.              }
47.          }
48.          results.clear();
49.          if(!hasMore){
50.              break;
51.          }
52.      }
53.      response = SumDTO.SumResponse.newBuilder().setSum(sum).build();
54.
55.  } catch (IOException ioe) {
56.      ResponseConverter.setControllerException(controller, ioe);
57.  } finally {
58.      if (scanner != null) {
59.          try {
60.              scanner.close();
61.          } catch (IOException ignored) {}
62.      }
63.  }
64.  done.run(response);
65. }
66.
67.
68. @Override public void start(CoprocessorEnvironment coprocessorEnvironment) throws IOException {
69.
70.     if (coprocessorEnvironment instanceof RegionCoprocessorEnvironment) {
71.         this.regionCoprocessorEnvironment = (RegionCoprocessorEnvironment) coprocessorEnvironment;
72.     } else {
73.         throw new CoprocessorException("Must be loaded on a table region!");
74.     }
75. }
76.
77.
78. @Override public void stop(CoprocessorEnvironment coprocessorEnvironment) throws IOException {
79.
80. }
81.
82.
83. @Override public Service getService() {
84.     return this;
85. }
86. }
```

代码清单 8-3 显示了如何在客户端调用该求和端点类型协处理器以得到订单总金额。

代码清单 8-3 求和端点类型协处理器测试程序

```
1.  package com.mt.hbase.chpt08.coprocessor;
2.
3.  import com.google.protobuf.ServiceException;
```

```
4.    import com.mt.hbase.connection.HBaseConnectionFactory;
5.    import com.mt.hbase.coprocessor.generated.SumDTO;
6.    import org.apache.hadoop.hbase.TableName;
7.    import org.apache.hadoop.hbase.client.Table;
8.    import org.apache.hadoop.hbase.client.coprocessor.Batch;
9.    import org.apache.hadoop.hbase.ipc.BlockingRpcCallback;
10.
11.   import java.io.IOException;
12.   import java.util.Map;
13.
14.   public class SumOrderEndpointTest {
15.
16.
17.       public static void main(String [] args) throws IOException {
18.
19.           TableName tableName = TableName.valueOf("s_order");
20.           Table table = HBaseConnectionFactory.getConnection().getTable(tableName);
21.
22.           final SumDTO.SumRequest request =
      SumDTO.SumRequest.newBuilder().setFamily("cf").setColumn("c")
23.               .build();
24.           try {
25.               Map<byte[], Long> results = table.coprocessorService(
      SumDTO.SumService.class, null, null,
26.                   new Batch.Call<SumDTO.SumService, Long>() {
27.                       @Override
28.                       public Long call(SumDTO.SumService aggregate) throws
      IOException {
29.                           BlockingRpcCallback<SumDTO.SumResponse> rpcCallback =
      new BlockingRpcCallback<SumDTO.SumResponse>();
30.                           aggregate.getSum(null, request, rpcCallback);
31.                           SumDTO.SumResponse response = rpcCallback.get();
32.                           return response.getSum();
33.                       }
34.                   });
35.               for (Long sum : results.values()) {
36.                   System.out.println("Sum = " + sum);
37.               }
38.           } catch (ServiceException e) {
39.               e.printStackTrace();
40.           } catch (Throwable e) {
41.               e.printStackTrace();
42.           }
43.       }
44.   }
```

到这里两种类型的协处理器都已经定义完成，接下来详细介绍如何装载和使用上面介绍的两种协处理器。

8.3 装载/卸载协处理器

无论是观察者类型协处理器还是端点类型协处理器，都是在分区服务器端执行，因此使用之前需要先将协处理器加载到 HRegionServer。

以下步骤描述了如何将前面编写的协处理器代码加载到分区服务器。

（1）将前面编写的协处理器代码打成 jar 包，例如使用 mvn 管理工程，使用 mvn package 将工程打包成 jar 文件，这里假设为 referencebook-1.0.jar。

（2）将打包好的 jar 文件上传至 HBASE_HOME/lib 目录。

（3）如果使用动态加载，也可以将 jar 文件上传至 HBase 所在集群的 HDFS，如代码清单 8-4 所示。

代码清单 8-4　上传文件到 HBase HDFS

```
hadoop fs -mkdir /hadoop
hadoop fs -put referencebook-1.0.jar hdfs://master1:9000/hadoop/referencebook-1.0.jar
```

上传成功后在使用之前需要装载协处理器，如果协处理器已经完成了其使命，也可以卸载已经装载的协处理器。

8.3.1 静态装载/卸载

静态装载/卸载通过修改配置项，重启 HRegionServer 进程实现，会影响到 HBase 集群所有的表。

1. 静态装载

静态装载需要修改 hbase-site.xml 配置文件以增加配置项，然后重启 HRegionServer，需要增加的配置项如代码清单 8-5 所示。

代码清单 8-5　协处理器静态装载配置项

```
<property>
    <name>hbase.coprocessor.region.classes</name>
    <value>com.mt.hbase.chpt08.coprocessor.endpoint.SumOrderEndpoint </value>
</property>
```

value 的值就是需要装载的自定义的协处理器类全名，name 有以下几种取值。

- hbase.coprocessor.region.classes：当需要装载的自定义的协处理器继承自 RegionObserver 接口或者 EndpointObserver 接口时。
- hbase.coprocessor.wal.classes：当需要装载的自定义的协处理器继承自

WALObserver 接口时。
- `hbase.coprocessor.master.classes`：当需要装载的自定义的协处理器继承自 `MasterObserver` 接口时。

如果要装载多个协处理器类，则类名需要以逗号分隔。HBase 会使用默认的类加载器加载配置中的这些类，因此需要将相应的 jar 文件上传到 `HBASE_HOME/lib`。

使用这种方式装载的协处理器将会作用在 HBase 所有表的全部分区上，因此静态装载的协处理器又被称为系统协处理器。在协处理器列表中第一个协处理器的优先级值为 `Coprocessor.Priority.SYSTEM`，之后的每个协处理器的优先级值将会按序加一（这意味着优先级会降低，优先级与数值的大小成反比），代码清单 8-6 列出了 HBase 定义的几种优先级。当调用配置的协处理器时，HBase 将会按照优先级顺序依次调用它们的回调方法。

代码清单 8-6　协处理器优先级定义

```
/** Highest installation priority */
int PRIORITY_HIGHEST = 0;
/** High (system) installation priority */
int PRIORITY_SYSTEM = Integer.MAX_VALUE / 4;
/** Default installation priority for user coprocessors */
int PRIORITY_USER = Integer.MAX_VALUE / 2;
/** Lowest installation priority */
int PRIORITY_LOWEST = Integer.MAX_VALUE;
```

2. 静态卸载

静态卸载就很容易了，移除增加到 hbase-site.xml 的配置项，然后重启 HRegionServer 进程即可。上传的 jar 包可以删除也可以保留。

8.3.2　动态装载 / 卸载

动态装载是针对表级别，优点是无须重启 HRegionServer 进程，但是由于是对表的模式层面的变动，因而需要将目标表离线。

1. 动态装载

动态装载协处理器有两种方式：通过 HBase shell 和通过 Java API。代码清单 8-7 演示了如何使用 HBase shell 为表 `s_behavior` 动态装载代码清单 8-1 定义的敏感词过滤协处理器。

代码清单 8-7　HBase shell 动态加载敏感词过滤协处理器

```
hbase(main):022:0> drop 's_behavior'
0 row(s) in 1.2320 seconds

hbase(main):023:0> create 's_behavior', {NAME => 'cf'}
```

```
0 row(s) in 1.2160 seconds

=> Hbase::Table - s_behavior
hbase(main):024:0> disable 's_behavior'
0 row(s) in 2.2280 seconds

hbase(main):025:0> alter 's_behavior', METHOD => 'table_att',
'Coprocessor'=>'hdfs://master1:9000/hadoop/referencebook-1.0.jar|
com.mt.hbase.chpt08.coprocessor.
KeyWordFilterRegionObserver| 12345| arg1=1,arg2=2'
Updating all regions with the new schema...
1/1 regions updated.
Done.
0 row(s) in 1.8850 seconds

hbase(main):026:0> enable 's_behavior'
0 row(s) in 1.2270 seconds

hbase(main):027:0> put 's_behavior','user1behavior1','cf:o','阿里巴巴this is order 1'
0 row(s) in 0.0130 seconds


hbase(main):029:0> scan 's_behavior'
ROW                         COLUMN+CELL
 user1behavior1             column=cf:o, timestamp=1509346549243, value=this
  ** is order 1
1 row(s) in 0.0100 seconds
```

从最后的 scan's_behavior'命令结果可以看到,敏感词已经被替换掉了。

代码清单 8-7 所示 HBase shell 的重点在如下这条命令:

```
alter 's_order', METHOD => 'table_att',
'Coprocessor'=>
'hdfs://master1:9000/hadoop/referencebook-1.0.jar|
com.mt.hbase.chpt08.coprocessor.KeyWordFilterRegionObserver| 12345| arg1=1,arg2=2'
```

这条命令重点在 Coprocessor 属性,该属性使用管道符"|"分隔成如下 4 个部分。

- 类文件路径:描述了协处理器所在的 jar 包的 HDFS 地址,支持通配符。
- 协处理器类名:需要装载的协处理器全类名。
- 优先级:一个整数,表示该协处理器执行的优先级。
- 参数(可选):这些参数对会被传递给要使用的协处理器的实现。

相对使用全局装载协处理器和使用 HBase shell 协处理器,Java API 提供了一个更加灵活装载协处理器的方法,在线上应用中可以根据业务需求随时随地装载所需要的协处理器,代码清单 8-8 演示了如何使用 Java API 装载自定义的求和端点类型协处理器。

代码清单 8-8 Java API 装载求和端点类型协处理器

```
1.   package com.mt.hbase.chpt8.coprocessor;
2.
```

```java
3.  import com.mt.hbase.connection.HBaseConnectionFactory;
4.  import org.apache.hadoop.hbase.Coprocessor;
5.  import org.apache.hadoop.hbase.HColumnDescriptor;
6.  import org.apache.hadoop.hbase.HTableDescriptor;
7.  import org.apache.hadoop.hbase.TableName;
8.  import org.apache.hadoop.hbase.client.Admin;
9.  import org.apache.hadoop.hbase.client.Connection;
10.
11. import java.io.IOException;
12.
13.
14. public class LoadCoprocessor {
15.
16.     public static void main(String[] args) throws IOException {
17.
18.         String path = "hdfs://master1:9000/hadoop/referencebook-1.0.jar";
19.         Connection connection = HBaseConnectionFactory.getConnection();
20.         Admin admin = connection.getAdmin();
21.
22.         TableName tableName = TableName.valueOf("s_behavior");
23.         admin.disableTable(tableName);
24.         HTableDescriptor hTableDescriptor = new HTableDescriptor(tableName);
25.         HColumnDescriptor cf = new HColumnDescriptor("cf");
26.         cf.setMaxVersions(2);
27.         hTableDescriptor.addFamily(cf);
28.
29.         hTableDescriptor.setValue("COPROCESSOR$1", path + "|"
30.                 + SumOrderEndpoint.class.getCanonicalName() + "|"
31.                 + Coprocessor.PRIORITY_USER);
32.         admin.modifyTable(tableName, hTableDescriptor);
33.         admin.enableTable(tableName);
34.     }
35. }
```

2. 动态卸载

HBase 同样提供了 HBase shell 和 Java API 两种方式来动态卸载协处理器，代码清单 8-9 和代码清单 8-10 分别演示了这两种方式。

代码清单 8-9　HBase shell 动态卸载敏感词过滤协处理器

```
hbase(main):017:0> disable 's_behavior'
0 row(s) in 2.2630 seconds

hbase(main):018:0> alter 's_behavior', METHOD => 'table_att_unset', NAME => 'coprocessor$1'
Updating all regions with the new schema...
1/1 regions updated.
Done.
0 row(s) in 1.8890 seconds
```

```
hbase(main):019:0> enable 's_behavior'
0 row(s) in 1.2340 seconds
```

代码清单 8-10　Java API 卸载求和端点类型协处理器

```
1.   package com.mt.hbase.chpt08.coprocessor;
2.
3.   import com.mt.hbase.connection.HBaseConnectionFactory;
4.   import org.apache.hadoop.hbase.Coprocessor;
5.   import org.apache.hadoop.hbase.HColumnDescriptor;
6.   import org.apache.hadoop.hbase.HTableDescriptor;
7.   import org.apache.hadoop.hbase.TableName;
8.   import org.apache.hadoop.hbase.client.Admin;
9.   import org.apache.hadoop.hbase.client.Connection;
10.
11.  import java.io.IOException;
12.
13.  public class LoadCoprocessor {
14.
15.      public static void main(String[] args) throws IOException {
16.
17.          String path = "hdfs://master1:9000/hadoop/referencebook-1.0.jar";
18.          Connection connection = HBaseConnectionFactory.getConnection();
19.          Admin admin = connection.getAdmin();
20.
21.          TableName tableName = TableName.valueOf("s_behavior");
22.          admin.disableTable(tableName);
23.          HTableDescriptor hTableDescriptor = new HTableDescriptor(tableName);
24.          HColumnDescriptor cf = new HColumnDescriptor("cf");
25.          cf.setMaxVersions(2);
26.          hTableDescriptor.addFamily(cf);
27.
28.          admin.modifyTable(tableName, hTableDescriptor);
29.          admin.enableTable(tableName);
30.      }
31.  }
```

协处理器为 HBase 提供了一些激动人心的特性，如以业务解耦的方式建立二级索引、复杂过滤器的实现等，通过将一些复杂的计算逻辑从客户端移动到服务端，减少了网络开销，从而获得很好的性能提升。

第 9 章

HBase 性能调优

在线的 OLTP 系统对响应时间要求非常高,想象一下当你在电商网站浏览某个商品或者下单支付的时候,如果系统几秒钟没有响应,你的耐心就可能被耗尽,进而使你放弃交易,因此当 HBase 为 OLTP 系统提供在线实时的数据存储时,响应时间以及吞吐量尤为重要。某一个配置项的不妥当可能直接造成线上 HBase 集群整体响应时间超时,然后应用服务器线程池耗尽,最终导致服务不可用,而一些简单的配置改动可能会让 HBase 集群性能提升数倍,因此 HBase 在线调优对 HBase 在企业生产环境的应用非常重要。

9.1 客户端调优

因为客户端调优仅影响单个客户端与 HBase 服务端交互的性能,所以调优风险也相对较小。

9.1.1 设置客户端写入缓存

如果业务能够容忍数据的丢失,如一些日志数据,那么客户端写入 HBase 表的时候可以采取批量缓存的方式,数据先缓存在客户端,当达到配置的阈值的时候再批量提交到服务端,注意如果客户端重启或者宕机,则这部分缓存的数据会丢失。

代码清单 9-1 演示了在表级别设置写入缓存。

代码清单 9-1 表级别设置写入缓存
```
1.   package com.mt.hbase.chpt09.client;
2.
3.   import com.mt.hbase.chpt05.rowkeydesign.RowKeyUtil;
```

```
4.   import com.mt.hbase.connection.HBaseConnectionFactory;
5.   import org.apache.hadoop.hbase.TableName;
6.   import org.apache.hadoop.hbase.client.HTable;
7.   import org.apache.hadoop.hbase.client.Put;
8.   import org.apache.hadoop.hbase.util.Bytes;
9.
10.  import java.io.IOException;
11.  import java.util.ArrayList;
12.  import java.util.List;
13.
14.  /**
15.   * Created by pengxu on 2018/2/20.
16.   */
17.  public class ClientBuffer {
18.
19.
20.      private static final String TABLE="s_behavior";
21.
22.      private static final String CF_PC="pc";
23.
24.      private static final String COLUMN_VIEW="v";
25.
26.      private static final long userId = 12345;
27.
28.      private static RowKeyUtil rowKeyUtil = new RowKeyUtil();
29.
30.      public static void main(String[] args) throws IOException,
    InterruptedException {
31.          List<Put> actions = new ArrayList<Put>();
32.          Put put = new Put(Bytes.toBytes(generateRowkey(userId,
    System.currentTimeMillis(),1)));
33.          put.addColumn(Bytes.toBytes(CF_PC), Bytes.toBytes(COLUMN_VIEW),
34.              Bytes.toBytes("1001"));
35.          actions.add(put);
36.          HTable table = (HTable)HBaseConnectionFactory.getConnection().
    getTable(TableName.valueOf(TABLE));
37.
38.          table.setAutoFlushTo(false);
39.          table.setWriteBufferSize(1024 * 1024 * 10);// 缓存大小 10M
40.
41.          Object[] results = new Object[actions.size()];
42.          table.batch(actions, results);
43.      }
44.
45.      private static String generateRowkey(long userId, long timestamp, long seqId){
46.          return rowKeyUtil.formatUserId(userId) +
    rowKeyUtil.formatTimeStamp(timestamp)+seqId;
47.      }
48.
49.  }
```

代码清单 9-2 则演示了在连接级别设置写入缓存。

代码清单 9-2　连接级别设置写入缓存
```
1.    private static synchronized void createConnection() {
2.
3.          Configuration configuration = HBaseConfiguration.create();
4.          configuration.set("hbase.client.pause", "100");
5.          // 缓存大小 10 MB
6.          configuration.set("hbase.client.write.buffer", "10485760");
7.          configuration.set("hbase.client.retries.number", "5");
8.          configuration.set("hbase.zookeeper.property.clientPort", "2181");
9.          configuration.set("hbase.client.scanner.timeout.period", "100000");
10.         configuration.set("hbase.rpc.timeout", "40000");
11.         configuration.set("hbase.zookeeper.quorum", "master1,master2,slave1");
12.
13.         try {
14.             connection = ConnectionFactory.createConnection(configuration);
15.         }catch(IOException ioException){
16.             throw new RuntimeException(ioException);
17.         }
18.    }
```

9.1.2　设置合适的扫描缓存

Scan 操作一般需要查询大量的数据，如果一次 RPC 请求就将所有数据都加载到客户端，则请求时间会比较长，同时由于数据量大，网络传输也容易出错，因此 HBase Scan API 提供了一个分批拉取数据然后缓存到客户端的功能。每次 `ResultScanner.next()` 被调用时，如果当前客户端扫描缓存数据为空，HBase 客户端就会去服务端拉取下一批数据，如果缓存值设置得过大，每次获取的数据过多，那么容易造成请求超时，甚至由于数据过大造成内存 `OutOfMemory` 异常；如果缓存值设置得过小，那么就增加了一些额外的 RPC 网络请求。因此一般会根据业务需求做一个平衡，设置一个最适合业务的值，如 1000 等，代码清单 9-3 演示了设置批量拉取数据的代码片段。

代码清单 9-3　设置合适的扫描缓存
```
1.    Scan scan = new Scan();
2.    scan.setCaching(1000);//表示一个 RPC 请求读取的数据条数
```

9.1.3　跳过 WAL 写入

WAL（Write-Ahead-Log）预写入日志，用作分区服务器异常恢复，在第 7 章有详细介绍，数据的写入操作需要等待 WAL 刷新写入到文件系统，因此对于一些能够容忍部分数据丢失的业务，如日志系统等，可以跳过 WAL 写入以提高写入速度，如代码清单 9-4 所示。

代码清单 9-4　跳过 WAL 写入
1. `Put.setWriteToWAL(false);`
2. `Delete.setWriteToWAL(false);`

9.1.4　设置重试次数与间隔

当 HBase 客户端请求在服务端出错并抛出异常后，如果抛出的异常不是 `DoNotRetryIOException` 类的子类，那么客户端会发起重试。客户端超时时间、重试的间隔与次数需要配置合理，否则容易造成分区服务器请求雪崩，进而导致应用服务器线程池线程耗尽，系统无法正常响应，以下两个配置项决定了重试的次数以及重试间隔。

- `hbase.client.pause`：重试的休眠时间系数。
- `hbase.client.retries.number`：最大重试次数，默认为 35，建议减少，如 5。

重试间隔为休眠时间系数乘以 A，其中 A=`RETRY_BACKOFF`[重试次数]，`RETRY_BACKOFF` 是一个常数数组，如代码清单 9-5 所示。

代码清单 9-5　跳过 WAL 写入
1. `public static final int [] RETRY_BACKOFF = {1, 2, 3, 5, 10, 20, 40, 100, 100, 100, 100, 200, 200};`

如果设置 `hbase.client.pause`=1000，`hbase.client.retries.number`=10，那么 10 次重试间隔为 1、2、3、5、10、20、40、100、100、100，单位秒。

如果重试次数超过了 `RETRY_BACKOFF` 数组大小，则 A=200（数组最后一个元素）。

HBase 1.2.6 源码里面描述了如下几种可重试的异常：

- `NotServingRegionException`；
- `RegionServerStoppedException`；
- `OutOfOrderScannerNextException`；
- `UnknownScannerException`；
- `ScannerResetException`。

9.1.5　选用合适的过滤器

Scan 请求通常需要扫描大量的数据行，过滤器可以用来在服务端过滤掉一部分不需要的数据，从而减少在服务端和客户端之前传输的数据量。下面列出了一些使用过滤器来提升 Scan 请求性能的手段。

- 组合使用 `KeyOnlyFilter`、`FirstKeyOnlyFilter`。`KeyOnlyFilter` 可以使得

服务端返回的数据量只包含行键，`FirstKeyOnlyFilter`可以减少服务端扫描的数据量，只需要扫描到每行的第一列。
- 避免使用包含大量`Filter`的`FilterList`。假如在用户行为日志管理系统中查询出商品 ID 从 1001 到 9999 的订单数据，此时使用包含多个`SingleColumnValueFilter`的`FilterList`可以满足需求，多个过滤器的关系为`Operator.MUST_PASS_ONE`（类似于 MySQL 的 `OR`），但是性能可能会非常低下，反而采用扫描出用户所有的订单数据到客户端过滤性能会更好。笔者试过在 7000 多的用户数据行，每行数据大小约 200 字节，使用 `SingleColumnValueFilter` 时，100 个以内使用 `FilterList` 的过滤操作速度会很快，一旦超过 100 个，则使用 `Scan` 设置起止行键扫描用户所有的数据到内存再过滤反而会更快。当然这与用户的数据量大小有关，并不是说 100 就是一个最优数字，不同业务可以通过实验得到不同的最优数字。
- 过滤器尽量使用字节比较器，因为 HBase 数据以字节形式存储。

9.2 服务端调优

服务端调优相比客户端调优更加重要也更加有效，当然调优风险也更大。由于不同类型的数据集有不同的访问模式，而且对服务如响应时长的要求也不一样，因此 HBase 官方给出了以下一些基本面上的建议，读者可能需要根据自己业务的不同而做出调整。
- 分区的大小在 10~50 GB，如果超过 100 GB，一旦发生拆分或者主压缩将对线上服务造成灾难性的打击。
- 单元格不要大于 10 MB，如果存在一些大的文件或者二进制的数据，则建议将数据存在 HDFS，然后将 HDFS 的路径保存在 HBase 单元格中。
- 一个表尽量只设计一个列族，一个表不应包含超过 3 个列族，因为 MemStore 的刷新与 StoreFile 的压缩是分区级别的，一个表所有的列族会被一起刷新或者压缩，而其中某些列族当前使用的 MemStore 的数据量可能很小，根本无须刷新，这样会导致一些无用的 IO。另外也有可能影响性能，假如一个表有 A 和 B 两个列族，列族 A 有 100 万行数据，而列族 B 有 1 亿行数据，那么列族 A 的数据可能跨多个分区甚至多个分区服务器，这样针对列族 A 的 Scan 操作会很低效。

9.2.1 建表 DDL 优化

建表 DDL 语句可以非常简单，只需指定表明与列族名，但是这条简单的建表语句隐藏了很多表属性，如代码清单 9-6 所示。

代码清单 9-6　建表语句

```
hbase(main):003:0> create 's_behavior',{NAME => 'cf'}
0 row(s) in 1.2280 seconds

=> Hbase::Table - s_behavior
hbase(main):004:0> desc 's_behavior'
Table s_behavior is ENABLED
s_behavior
COLUMN FAMILIES DESCRIPTION
{NAME => 'cf', DATA_BLOCK_ENCODING => 'NONE', BLOOMFILTER =>
'ROW', REPLICATION_SCOPE => '0', VERSIONS => '1', COMPRESSION =>
'NONE', MIN_VERSIONS => '0', TTL => 'FOREVER', KEEP_DELETED_CELLS =>
'FALSE', BLOCKSIZE => '65536', IN_MEMORY =>
'false', BLOCKCACHE => 'true'}
1 row(s) in 0.0240 seconds
```

从表的描述可知，HBase 赋予了该表很多默认属性，下面介绍如何通过这些属性来优化表性能。

1. 使用数据块编码

DATA_BLOCK_ENCODING 表示针对行键使用的数据块编码格式。常用的一种数据块编码格式为 PREFIX，当开启 PREFIX 编码后 HBase 数据存储文件中会添加一个保存着当前行行键与上一行行键具有的相同前缀字符的列。图 9-1 描述了一个没有启用数据编码的表数据，图 9-2 描述了同一个启用了数据编码的表数据。

键长	值长	键	值
41	…	54321000000000000000092233705133122008861	1002
41	…	54321000000000000000092233705133122008862	1004
41	…	54321000000000000000092233705133122008863	1009
41	…	54321000000000000000092233705133122009210	1001

图 9-1　表数据：不启用数据库编码

键长	值长	前缀长	键	值
41	…	0	54321000000000000000092233705133122008861	1002
		40	2	1004
		40	3	1009
41	…	37	9210	1001

图 9-2　表数据：启用数据库编码

除了 PREFIX 编码格式外，HBase 还支持以下几种数据块编码格式。

- DIFF：对 PREFIX 的一种扩展，将 PREFIX 对键前缀的缩略扩展到值与时间戳（存偏移量而不是相同前缀字符数量）。DIFF 对读写会有比较大的影响，但是可以缓存更多数据，一般不启用。
- FAST_DIFF：一种对 DIFF 更快的实现，与 DIFF 区别不大。
- PREFIX_TREE：前缀树，这种编码对内存使用率的提高（压缩了数据，相同内存可以缓存更多数据）与其他几种编码格式差别不大。前缀树以牺牲编码速度换取更快的随机读取速度，前缀树比较适合缓存命中率比较高的应用，它引入了一个tree 的字段，这个字段包含了一系列对一行单元格的偏移量和引用。在没有PREFIX_TREE 之前，无论是否开启数据块编码，表数据都是按照 KeyValue 的排序，将其一条条顺序地写入数据块中，因此从数据块中查找数据，只能采用遍历的方式。开启了 PREFIX_TREE 后，数据块中的数据存储不再是简单地将 KeyValue 按照顺序进行堆压，而是按照特有的方式进行组织，可以在解析时生成前缀树，并且树节点的儿子是排序的，因此从数据块中查找数据，其效率至少在二分查找法以上。图 9-3 给出了 PREFIX_TREE 作者提供的几种编码查找性能的对比。

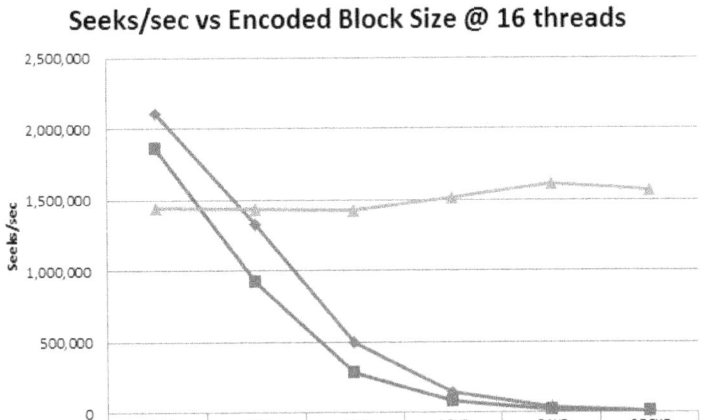
Seeks/sec vs Encoded Block Size @ 16 threads
Seeks/sec
2,500,000
2,000,000
1,500,000
1,000,000
500,000
0

	256B	1KB	4KB	16KB	64KB	256KB
NONE	2,109,706	1,331,488	493,733	138,557	38,036	9,726
PREFIX	1,863,481	923,550	283,914	77,858	20,630	5,248
TRIE	1,446,817	1,429,824	1,425,903	1,515,525	1,614,555	1,567,727

图 9-3　数据块编码方式性能对比（图片源自网络）

下面是对这个性能对比的几点说明。

- 作者列出了数据块大小从 256 B 到 256 KB 之间的几种性能对比，但是没有指出存储在数据库中的单个键值对的平均大小。
- NONE 指的是不启用数据块编码，PREFIX 指的是启用 PREFIX 编码格式，TRIE 是指启用 PREFIX_TREE 编码格式。
- 前缀树压缩算法在不同大小的数据块下查找性能都很稳定，而 NONE 和 PREFIX 因为是用遍历的方式查找数据，所以查找性能随着数据块增大而直线下降。在建表默认的 64 KB 的数据块情况下，PREFIX_TREE 性能要比另外两种高 40 多倍。
- NONE 比 PREFIX 算法性能高，是由于 PREFIX 算法需要解压。

笔者在生产环境下对数据量超过 30 TB 的表使用 PREFIX_TREE 编码时，绝大部分情况下运行正常，查询性能也确实有所提高，但是会偶现编码线程占用 CPU 资源很高的情况。

2. 使用布隆过滤器

BLOOMFILTER 表示使用布隆过滤器，布隆过滤器可以用来提高随机读的性能。HBase 支持 ROW 与 ROWCOL 类型的布隆过滤器，ROWCOL 只对指定列的随机读（Get）有效，如果应用中的随机读没有指定读哪个列限定符，那么设置 ROWCOL 是没有效果的，这种场景就应该使用 ROW。

布隆过滤器的原理是，当一个元素被加入集合时，通过 K 个散列函数将这个元素映射成一个位数组中的 K 个点，把它们的值置为 1。检索时，我们只要看这些点的值是不是都是 1 就（大约）知道集合中有没有它了：如果这些点中有任何一个 0，则被检元素一定不在；如果都是 1，则被检元素很可能在。

布隆过滤器的数据存在 StoreFile 的元数据中，开启布隆过滤器会有一定的存储以及内存开销，在生产环境中观察到一个开启了 snappy 压缩的表，平均 20 GB 左右的 Store（4~5 个 StoreFile），布隆过滤器占用的空间在 80 MB 左右，因此这个开销相对还是可以接受的。

3. 开启数据压缩

HBase 支持多种数据压缩方式用来减少存储到 HDFS 的文件大小以节约磁盘空间。常用的压缩算法包括：GZIP、Bzip2、LZO、SNAPPY，其中 LZO 与 SNAPPY 由于版权问题，需要在操作系统安装原生库才可以支持，例如 centOS 安装 SNAPPY：

```
yum install snappy snappy-devel:
```

图 9-4 描述了这些压缩算法的压缩比例与速度对比，从对比图可以看出，压缩比例越高，速度就越慢。相对来说 SNAPPY 能够提供比较均衡的性能，压缩比例尚可，吞吐量高，因此推荐使用。

Algorithm	% remaining	Encoding	Decoding
GZIP	13.4%	21 MB/s	118 MB/s
LZO	20.5%	135 MB/s	410 MB/s
Snappy	22.2%	172 MB/s	409 MB/s

图 9-4　HBase 不同压缩方式压缩比例与速度对比

压缩/解压是一个 CPU 型操作，因此启用 HBase 压缩会导致 CPU 使用率上升。数据压缩在写入 MemStore 之后，MemStore 刷新输出到磁盘之前，对写性能影响不大，但是读取数据时需要将数据块解压后才能读取以及放入缓存，因此对读性能其实有负面影响。

开启压缩很简单，只需以下两步。

（1）在 hadoop 配置文件 core-site.xml 中添加如代码清单 9-7 所示配置。

代码清单 9-7 添加 Hadoop 支持编码格式

```xml
<property>
    <name>io.compression.codecs</name>
    <value>
        org.apache.hadoop.io.compress.GzipCodec,
        org.apache.hadoop.io.compress.DefaultCodec,
        com.hadoop.compression.lzo.LzoCodec,
        com.hadoop.compression.lzo.LzopCodec,
        org.apache.hadoop.io.compress.BZip2Codec,
        org.apache.hadoop.io.compress.SnappyCodec
    </value>
</property>
```

（2）建表 DDL 增加属性：

```
COMPRESSION => 'SNAPPY'
```

4. 设置合理的数据块大小

BLOCKSIZE 属性决定了 HBase 读取数据的最小块大小。为了提升性能，理想情况下每次查询所需要扫描的数据都能够放到一个数据块，或者是数据块的整数倍。例如，用户的行为数据、同一个用户的所有数据能够集中地放在几个或者多个数据块当中。`HFile.java` 源码中有一段注释：我们推荐将数据块的大小设置为 8 KB～1 MB。大的数据块比较适合顺序查询（如 Scan），但不适合随机查询（因为需要解压缩一个大的数据块）。小的数据块适合随机查询，但是需要更多的内存来保存数据块的索引（data index），而且创建文件的时候也可能比较慢，因为在每个数据块的结尾我们都要把压缩的数据流 Flush 到文件中去（引起更多的 Flush 操作）。此外，由于压缩编解码器还需要一定的缓存，因此最小的数据块大小应该在 20～30 KB，默认的数据块大小为 64 KB，如下 DDL 建表属性设置数据块大小为 128 KB：

```
BLOCKSIZE => '131072'
```

一般情况下建议根据数据行的大小以及业务使用的模式来设置合理的块大小，默认的 64 KB 是在随机读与顺序读之间比较平衡的一个设置。

5. 预先分区

预分区的好处是可以负载均衡以充分利用集群中分区服务器的能力，最终提升整体的性能。每个表在设计时应该结合表的大小以及分区服务器的数量，来决定分区的多少，一般一个分区的 Store File 的大小可以在 10～50 GB，太大的 StoreFile 压缩或者拆分会对集群性能造成灾难性的影响。一个分区服务器上分区的数量也最好不要太多，假如一个表预计存储数据 1 TB 左右，那么比较合适的预分区方案是将表分为 50～100 个分区，这样最终每个分区的 StoreFile 大小在 10～20 GB。当然，如果集群中分区服务器的数量非常大，例如上百台，那么将表分为更多的分区（超过集群分区服务器数量的 1 倍）会是一个更优的方案。

以下 DDL 建表代码片段可以用来将分区预先分为 10 个区：

SPLITS =>['1' ,'2','3' ,'4' ,'5','6','7','8','9']

9.2.2 禁止分区自动拆分与压缩

默认情况下 HBase 自动管理分区的拆分（split）和主压缩（major compact，又称为主合并），这样可以减少一些运维工作。在行键设计合理的情况下，分区基本上都是均衡地增长的，如果采用自动拆分与主压缩，很容易造成拆分和压缩风暴，在生产环境中一个 30 GB 左右的 Store 主压缩时间可以超过 20 分钟。如果是离线集群，则可能不会太关注拆分和压缩对集群性能的影响，但是对实时在线集群来说，这个影响可能就无法接受了，尤其是主压缩需要合并 StoreFile 文件、清理过期数据、清理删除状态的数据以及将数据本地化，这对网络（对分区服务器的网络带宽占用可以超过 1 Gbit/s）和磁盘 IO 都有很大的压力，因此对在线集群一般需要禁止自动拆分和主压缩，然后在合适的时候进行手动拆分和手动主压缩（一般在业务非高峰期做拆分和压缩），同时将拆分和主压缩操作错开，将这些操作产生的网络与磁盘 IO 负载平衡，最小化对线上服务的影响。

手动拆分还有一个优势是线上分区的名称与数量不会变化，这样一些自动化的监控程序或者调试脚本可以很好地工作，而自动化分区拆分会对分区重命名、监控或者运维带来麻烦。

1. 禁止自动拆分

参数 `hbase.regionserver.regionSplitLimit` 控制 HBase 表最大的分区数量，超过则不能进行自动拆分。建议在建表的时候先预分区，然后在需要的时候（例如某个分区的 StoreFile 超过 50 GB）选择集群空闲时间执行手动拆分。代码清单 9-8 列出了为了禁止分区自动拆分需要添加到配置文件 `hbase-site.xml` 的配置项。

代码清单 9-8　禁止分区自动拆分 hbase-site.xml

```
<property>
    <name>hbase.regionserver.regionSplitLimit</name>
    <value>1</value>
</property>
```

2. 禁止自动主压缩

下列配置项与压缩相关。

- `hbase.hregion.majorcompaction`：该参数控制主压缩间隔的时间，默认为

604 800 000 ms，即 7 天，设置为 0 表示禁止自动主压缩。
- `hbase.hstore.blockingStoreFiles`：当 Store 的 StoreFile 数量超过该参数配置的值时，需要在刷新 MemStore 前先进行拆分或者压缩，除非等待超过 `hbase.hstore.blockingWaitTime` 配置的时间，默认的配置时间为 90 000 ms，因此需要适量调大该参数，以免 Memstore 刷新被阻塞，进而影响写入操作，导致整个分区服务器服务异常。
- `hbase.hstore.compactionThreshold`：当 Store 的 StoreFile 数量大于等于该参数配置的值时，可能会触发压缩。默认值为 3，如果配置值过大，可以推迟触发压缩的时间，但是会造成 Store 的 StoreFile 数量过大，影响查询的性能，一般设置为 5 以内。

代码清单 9-9 列出了为了禁止主压缩需要添加到配置文件 hbase-site.xml 的配置项。

代码清单 9-9　禁止分区自动主压缩 hbase-site.xml

```xml
<property>
    <name>hbase.hregion.majorcompaction</name>
    <value>0</value>
</property>
<property>
    <name>hbase.hstore.blockingStoreFiles</name>
    <value>500</value>
</property>
<property>
    <name>hbase.hstore.compactionThreshold</name>
    <value>5</value>
</property>
```

9.2.3　开启机柜感知

机柜感知实际上是属于 Hadoop HDFS 的一个优化。HDFS 为了保证数据的安全，数据文件默认会在 HDFS 上保存 3 个副本，存储策略是本地一份，另外一个机柜 A 上的机器 X 存一份，机柜 A 上的另外一台机器 Y 存一份。这样一方面能够保证数据能够就近读取，另一方面即使某个机柜上面的所有机器出现断电或者网络异常，HDFS 也仍然能够在另外一个机柜上找到数据的备份。那么 Hadoop 是怎样判断两台不同机器是否在同一个机柜呢？

这就需要配置机柜感知了，默认情况下机柜感知是关闭的，集群中所有的机器都在一个默认的机柜下，这样可能会造成网络流量的浪费。假设集群中总共有 6 台物理机，机器 A、B、C、D、E 和 F 分属于两个机柜，机器 A、B 和 C 属于机柜 M，机器 D、E 和 F 属于机柜 N，Hadoop 配置的数据副本数为 3。在没有配置机柜感知的情况下，当机器 A 有数据写入时，第一个数据副本存放在 A，另外两个副本可能存放在 B、C、D、E 和 F 中的任意两台机器，有可能第二个副本存放在机器 E，第三个副本存放在机器 B，这样数据的流向是 A→E→B，也就是说数据从机柜 M 流向机柜 N，然后又流回 M，显然从 N 流回 M 是不必要的。

下面步骤描述了如何开启机柜感知。

（1）增加 `core-site.xml` 配置项。在 Hadoop 的配置文件 `core-site.xml` 中增加代码清单 9-10 所示配置项，该配置项指定了配置机柜感知的脚本文件路径。

代码清单 9-10　配置机柜感知脚本文件路径 core-site.xml

```xml
<property>
    <name>topology.script.file.name</name>
    <value>/data/hadoop-2.6.5/etc/hadoop/topology.sh</value>
</property>
```

（2）增加 `topology.sh` 机柜感知脚本。`topology.sh` 脚本如代码清单 9-11 所示，该脚本需要读取一个配置文件 `topology.data` 来录入机柜信息。

代码清单 9-11　机柜感知脚本 topology.sh

```bash
#!/bin/bash
HADOOP_CONF=/data/hadoop-2.6.5/etc/hadoop
while [ $# -gt 0 ] ; do
  nodeArg=$1
  exec<${HADOOP_CONF}/topology.data
  result=""
  while read line ; do
    ar=( $line )
    if [ "${ar[0]}" = "$nodeArg" ]||[ "${ar[1]}" = "$nodeArg" ]; then
      result="${ar[2]}"
    fi
  done
  shift
  if [ -z "$result" ] ; then
    echo -n "/default-rack"
  else
    echo -n "$result"
  fi
done
```

（3）增加 `topology.data` 机柜配置信息。文件 `topology.data` 内容如代码清单 9-12 所示，该文件每行描述了一个 Hadoop 机器节点的机房机柜信息，格式为 "IP+主机名+/机房/机柜"，如下配置表示一个 6 台机器的集群，其中 master1、slave1 和 slave3 属于同一个机柜 rack1，另外 3 台属于机柜 rack2。

代码清单 9-12　机柜配置信息 topology.data

```
192.168.172.1 master1 /dc1/rack1
192.168.172.2 master2 /dc1/rack2
192.168.172.3 slave1 /dc1/rack1
```

```
192.168.172.4 slave2 /dc1/rack2
192.168.172.5 slave3 /dc1/rack1
192.168.172.6 slave4 /dc1/rack2
```

（4）重启 NameNode 使机柜感知生效。开启机柜感知后需要重启 NameNode，重启之后执行代码清单 9-13 所示命令即可查看配置是否生效。

代码清单 9-13　查看机柜配置信息

```
./hadoop dfsadmin -printTopology
Rack: /dc1/rack1
   192.168.172.1:50010 (master1)
   192.168.172.3:50010 (slave1)
   192.168.172.5:50010 (slave3)

Rack: /dc1/rack2
   192.168.172.2:50010 (master2)
   192.168.172.4:50010 (slave2)
   192.168.172.6:50010 (slave4)
```

如果新加机器节点到集群，则只需将新节点配置添加到 `topology.data`，然后调用如下命令即可使得机柜感知在新增节点生效：

```
./hadoop dfsadmin -refreshNodes
```

> **注意**　注意开启机柜感知也需要运维在机器物理摆放上配合，一般建议多个机柜，每个机柜上的机器比较均衡。假如一个集群的 6 台机器放在 2 个机柜 M 和 N 中，其中 M 放置 1 台机器，N 放置 5 台机器，那么 M 机柜上的一台机器肯定会成为瓶颈，读者可以自行思考一下原因。

9.2.4　开启 Short Circuit Local Reads

HBase 集群中的每个节点上一般都会同时部署 Hadoop DataNode 以及 HBase 分区服务器。HBase 分区服务器在主压缩（major compact）过程中会将多个小的 StoreFile 合并成一个大的 StoreFile，同时这个大的 StoreFile 会存放在负责该分区的分区服务器，这就是分区的本地化。

移动"计算"比移动"数据"容易。如果读取数据的 Hadoop DFSClient 与 DataNode 在同一个节点，则称之为"本地读"，与之相对的是"远程读"（remote read），也就是 Hadoop DFSClient 与 DataNode 不在同一个节点，那么读取数据就需要一个 RPC 调用。默认情况下不管本地读还是远程读，其实都需要经过一层 DataNode 的 RPC 转调，当开启 Short Circuit Local Reads 配置后，相当于 Hadoop DFSClient 直接读取本地文件，而无须经过 DataNode 的中转。

Short Circuit Local Reads 用到了 Unix Domain Socket，它是一种进程间的通信方式，使得同一台机器上的两个进程能以 Socket 的方式通信。它带来的另一大好处是，利用它两个进程间除可以传递普通数据外，还可以传递文件描述符。

假设机器上的两个用户 A 和 B，A 拥有访问某个文件的权限而 B 没有，而 B 又需要访问这个文件。借助 Unix Domain Socket，可以让 A 打开文件得到一个文件描述符，然后把文件描述符传递给 B，B 就能读取文件里面的内容了，即使 B 没有相应的权限。在 HDFS 的场景里面，A 就是 DataNode，B 就是 DFSClient，需要读取的文件就是 DataNode 数据目录中的某个文件。

添加代码清单 9-14 所示的代码到 Hadoop 配置文件 hdfs-site.xml 即可开启 Short Circuit Local Reads。注意需要重启 DataNode 和分区服务器后才会生效，并且需要创建一个空文件 /data/hadoop-2.6.5/dn_socket 用来进程间通信。

代码清单 9-14　开启 Short Circuit Local Reads hdfs-site.xml

```xml
<property>
    <name>dfs.client.read.shortcircuit</name>
    <value>true</value>
</property>
<property>
    <name>dfs.domain.socket.path</name>
    <value>/data/hadoop-2.6.5/dn_socket</value>
</property>
```

9.2.5　开启补偿重试读

HBase 数据文件基于 Hadoop HDFS 存储，为了保证数据的安全可靠，一般会存储多个备份，默认是 3 个。当开启了 Short Circuit Local Reads 后，HBase 读取数据会优先从本地读取，某些情况下由于本地磁盘或者网络问题可能会导致短时间内的本地读失败，为了应对这些情况，HBase 社区对这种情况提出了补偿重试读（hedged read）。

开启补偿重试机制后，当客户端发起一个本地读时，如果超过配置的时间还没返回，客户端就会向数据副本所在的其他 DataNode 发送相同的数据请求。哪一个请求先返回，另一个就会被丢弃。

将代码清单 9-15 所示的配置项添加到 `hbase-site.xml` 文件即可开启 HBase 补偿重试读。

代码清单 9-15　开启补偿重试读 hbase-site.xml

```xml
<property>
  <name>dfs.client.hedged.read.threadpool.size</name>
  <value>20</value>    <!--20个线程-->
```

```
</property>
<property>
  <name>dfs.client.hedged.read.threshold.millis</name>
  <value>5000</value>   <!--5000 ms -->
</property>
```

两个配置项解释如下。

- `dfs.client.hedged.read.threadpool.size`：并发补偿重试读的线程池大小。
- `dfs.client.hedged.read.threshold.millis`：补偿重试读开始前等待的时间，即如果一个请求在该配置的时间内还未返回，则发起重试读。

9.2.6　JVM 内存调优

当一个在线应用的访问压力较小的时候，程序性能通常能够达到预期，此时应用管理员都会用一套通用的 JVM 启动参数来应对这些应用。一旦每秒请求（QPS）达到一定的数量级，应用的响应时间（RT）通常会因为各种资源瓶颈而直线上升，JVM GC 时间就是一个影响 RT 的重要因素。

为了更好地阅读本节内容，读者需要对 JVM GC 有一定了解，包括了解 GC 的工作原理和常用的 GC 算法、理解新生代、老生代等术语。

Oracle 公司的 Hotspot 虚拟机提供了两个并发程度很高的 JVM 垃圾回收。

- Concurrent Mark Sweep Collector（CMS）："分代回收"算法中老生代的一种回收算法，适用于对停顿时间要求较短、可以为 GC 线程共享 CPU 资源的应用，通常与年轻代回收算法 "Parallel New Collector" 一起使用。
- Garbage-First Garbage Collector（G1）：JDK7u4 版本发行时被正式推出，其设计目标是用来替代 CMS。G1 能够更好地预测垃圾回收停顿时间，同时完成高吞吐量的目标，与 CMS 相比具有多个优点，例如能够进行内存整理，不会产生很多内存碎片；新生代与老生代分区不再固定，内存上使用更为灵活；在停顿时间上加了预测机制，用户可以指定停顿时间以免应用雪崩。

目前实际工作中大多数应用系统都在使用 CMS，但是 G1 是未来 JDK 垃圾回收器算法的方向。

JVM GC 调优的目标有以下两个。

- 减少 Minor GC 时间，因为 Minor GC 并发复制会使得 JVM 处于 STW（stop-the-word）状态，整个应用都暂停而无法响应。
- 减少 Full GC 次数，每次时间越少越好，因为除了标记阶段的 STW，Full GC 会产生大量内存碎片（使用 CMS 垃圾回收算法时），并且如果单次 GC 时间过长超过心跳阈值，则会导致 HBase 分区服务器被 ZooKeeper 认为已经死亡而从集群中移除。

1. CMS 垃圾回收器

CMS 发展到现在已经非常成熟了，读者可以轻松从各种资料中找到针对不同内存配置机器的推荐 JVM 配置参数。表 9-1 列出了与 CMS GC 相关的一些重要配置参数，实际应用的 JVM 调优通常会针对这些参数做适当调整，最终得到一个最适合自己业务的配置。

表 9-1 CMS GC 参数表

参数	解释
-XX:+UseConcMarkSweepGC	老年代启用 CMS 收集器
-XX:UseParNewGC	年轻代启用并行回收算法，通常与 CMS 一起启用
-Xmx8G	指定 JVM 堆内存最大可用值为 8 GB
-Xms8G	指定 JVM 堆内存的初始化大小，为了避免 JVM 堆扩容带来的开销，该值一般配置与 -Xmx 一致
-Xmn1G	指定 JVM 堆分配给新生代的大小为 1 GB，此参数对 JVM GC 性能影响很大，因为新生代的垃圾回收会 Stop-The-World
-XX:MaxDirectMemorySize=4G	指定 JVM 可以使用的最大堆外内存大小，当 HBase 启用 Bucket 缓存时（建议使用），该堆外内存配置的值必须大于 Bucket 缓存配置的可用内存大小
-Xss256K	指定分配给每个线程的栈大小为 256 KB，该值一般配置为 256 KB~1 MB
-XX:SurvivorRatio=3	指定年轻代 Eden 区与 Survivor 区内存大小比为 3，该值默认为 8，该值会影响新生代内存回收的开始时间
-XX:MaxTenuringThreshold=8	指定年轻代对象经过 8 次 GC 后可以晋升到老年代，该值默认为 15
-XX:ParallelGCThreads=5	指定年轻代的并行收集线程数为 5 个，如果机器 CPU 核心数目小于 8，则可以设置为 CPU 核心数，如果 CPU 核心数目大于 8，则建议设置为 CPU 核心数的 1/4 左右
-XX:ConcGCThreads=3	定义 CMS 并发过程线程数为 3 个，配置更多的线程数（当然不能超过机器 CPU 核心数）会加快并发 CMS 过程，但也会带来额外的同步开销。对于不同应用程序，需要在真实生产环境中通过真实的测试来确认增加 CMS 线程是否能够提升 GC 性能，如果该参数未配置，JVM 会根据并行收集器中的 -XX:ParallelGCThreads 参数的值来计算出默认的并行 CMS 线程数（ConcGCThreads = (ParallelGC-Threads + 3)/4）

参数	解释
-XX:CMSInitiatingOccupancyFraction=70	程序运行过程中会不停地创建新对象，需要新分配内存，因此不能到堆占满之后再进行内存回收，该参数指定第一次 CMS 垃圾回收启动的时机为堆内存使用达到 70%时，之后 JVM 会根据运行时对象分配与释放的统计来决定 CMS 垃圾回收周期
-XX:+UseCMSInitiatingOccupancyOnly	设置 JVM 不根据运行时对象的分配与释放统计来决定 CMS 垃圾回收周期，而根据-XX:CMSInitiatingOccupancyFraction 配置的值来决定每一次垃圾回收。注意配置该参数一定需要经过生产环境严格的验证，否则通常 JVM 基于运行时的统计能够做出更好的垃圾回收决策
-XX:+UseCMSCompactAtFullCollection	指定在 Full GC 的时候对老年代内存进行压缩，因为 CMS 是不移动内存数据的，回收后通常会产生内存碎片，该参数会影响性能，但是会消除碎片提供内存使用率
-XX:CMSFullGCsBeforeCompaction=3	指定 3 次 Full GC 后执行内存压缩
-XX:+DisableExplicitGC	指定 JVM 忽略系统调用的 GC（不限垃圾回收器类型）

根据上述参数说明，以 64 GB 内存 24 核机器为例，可以得到一个比较常用的 JVM 参数配置，如代码清单 9-16 所示。

代码清单 9-16　常用 JVM 参数配置

```
-Xss256K -Xmx16G -Xms16G -Xmn2G -XX:MaxDirectMemorySize=24g -XX:SurvivorRatio=3
-XX:+UseParNewGC -XX:+UseConcMarkSweepGC -XX:MaxTenuringThreshold=15
-XX:CMSInitiatingOccupancyFraction=70 -XX:+UseCMSCompactAtFullCollection
-XX:+UseCMSInitiatingOccupancyOnly -XX:+DisableExplicitGC
-XX:+HeapDumpOnOutOfMemoryError -XX:ParallelGCThreads=15 -XX:ConcGCThreads=4
```

接下来以表 9-2 所示某线上集群节点为例，通过调整如下关键参数，观察生产环境应用 GC 的次数与时间以得出应用的最佳 JVM 参数配置。

- -Xmn：增大参数值可以降低 Minor GC 频率，但是会增加单次 GC 时间；减小参数值会加快 Minor GC 频率，同时加快对象晋升到老年代的速度，潜在增加 Full GC 的概率。
- -XX:SurvivorRatio：该参数越大表示 Survivor 区越小，可能导致 Minor GC 时存活的对象没有达到配置的 `MaxTenuringThreshold` 值就直接进入老年代，增加了老年代 GC 的概率；参数越小则表示 Survior 区越大，这样长寿对象可以在年轻代待到达到 `MaxTenuringThreshold` 阈值时才进入老生代，这样也会增加这些对象在 Survior 区来回复制的次数与时间，增加了 Minor GC 时长。

表 9-2　CMS GC 参数表

参数	硬件	HBase 分区服务器 JVM 配置

Slave3	32 核心 CPU,
	128 GB 内存
		`-Xmx24g -Xms24g -Xmn4g -XX:MaxDirectMemorySize=26g`
		`-XX:SurvivorRatio=3 -XX:+UseParNewGC -XX:+UseConcMarkSweepGC`
		`-XX:MaxTenuringThreshold=8`

Slave7	24 核心 CPU,
	128 GB 内存
		`-Xmx24g -Xms24g -Xmn4g -XX:MaxDirectMemorySize=26g`
		`-XX:SurvivorRatio=3 -XX:+UseParNewGC -XX:+UseConcMarkSweepGC`
		`-XX:MaxTenuringThreshold=8`

Slave8	32 核心 CPU,
	128 GB 内存
		`-Xmx24g -Xms24g -Xmn2g -XX:MaxDirectMemorySize=26g`
		`-XX:SurvivorRatio=3 -XX:+UseParNewGC -XX:+UseConcMarkSweepGC`
		`-XX:MaxTenuringThreshold=8`

运行一段时间后，使用 `jstat` 命令 `jstat -gcutil pid` 统计 3 个节点 GC 的耗时，结果如图 9-5 所示。

机器	S0	S1	E	O	P	YGC	YGCT	FGC	FGCT	GCT	运行时长（秒）
slave3	14.43	0	66	78.53	36.43	33864	813.36	30	3.726	817.085	61108.336
slave8	13.1	16.52	0	56.27	35.23	99941	1924.864	44	4.078	2329.212	88729.063
slave7	0	10.91	78.47	62.02	36.24	66077	1981.28	14	0.352	1512.127	27620.727

机器	平均YGC时长（毫秒）	平均FGC时间（毫秒）	每小时YGC次数	每小时FGC次数	每小时YGC时长（毫秒）	每小时FGC时长（毫秒）
slave3	24.01	124.2	1995	1.77	47920	219.5
slave8	19.26	92.68	4055	1.79	78099	165.5
slave7	36.99	141.52	1979	1.94	73226	274.5

图 9-5　CMS GC 耗时对比

除此之外，线上运行过程中，slave3 与 slave7 的 HBase 分区服务器日志文件经常出现如代码清单 9-17 所示的 JVM 耗时日志，从日志可以看出由于新生代 GC（`ParNew`）造成了 JVM 的停顿，总时长 1830 ms。

代码清单 9-17　JVM 停顿日志

```
2018-06-29 12:39:38,644 INFO  [JvmPauseMonitor] util.JvmPauseMonitor: Detected
pause in JVM or host machine (eg GC): pause of approximately 1486ms
GC pool 'ParNew' had collection(s): count=1 time=1830ms
```

综上所述，对 CMS GC 调优测试结果与预期表现一致，可以得到如下结论（不同业务不同集群可能有不同结果）。

- 较小的 `-Xmn` 可以减少 YGC 时间，但是会增加 YGC 次数，较大的 `-Xmn` 配置可能使得 Minor GC 时间较长进而有可能导致 HBase 分区服务器被 ZooKeeper 认为宕机，因此一般 32 GB 以下的堆内存建议 `-Xmn` 配置不超过 2 GB。

- CPU 核心数越多，新生代 GC 速度越快，耗时越短。
- 该 HBase 线上集群 "朝生夕死" 的对象较多，因此 YGC 次数增加并没有导致晋升到老生代的对象增加，从而导致 Full GC 次数增加。

2. G1 垃圾回收器

G1 垃圾回收器相对 CMS GC 最大的变化是将堆内存划分为多个大小相等的堆分区（也称为 region，这里的 region 和 HBase 中的 region 是两个概念），其垃圾回收过程与 CMS GC 类似，回收的时候会将堆分区中存活的对象转移，解决了 CMS GC 内存碎片的问题。

表 9-3 列出了与 G1 相关的一些重要配置参数。

表 9-3 G1 GC 参数表

参数	解释
-XX:+UseG1GC	开启 G1 GC
-XX:G1HeapRegionSize=4M	设置 G1 堆分区大小为 4 MB，最小值为 1 MB，最大为 32 MB，必须为 2 的幂次方，目标是产生不超过 2048 个分区
-XX:MaxGCPauseMillis=300	设置 GC 暂停的期望时间为 300 ms，默认值是 200 ms
-XX:G1NewSizePercent	新生代最小值，默认值 5%
-XX:G1MaxNewSizePercent	新生代最大值，默认值 60%
-Xmx8G	指定 JVM 堆内存最大可用值为 8 GB
-Xms8G	指定 JVM 堆内存的初始化大小，为了避免 JVM 堆扩容带来的开销，该值一般配置与 -Xmx 一致
-XX:MaxDirectMemorySize=4G	指定 JVM 可以使用的最大堆外内存大小，当 HBase 启用 Bucket 缓存时（建议使用），该堆外内存配置的值必须大于 Bucket 缓存配置的可用内存大小
-Xss256K	指定分配给每个线程的栈大小为 256 KB，该值一般配置为 256 KB～1 MB
-XX:SurvivorRatio=3	指定年轻代 Eden 区与 Survivor 区内存大小比为 3，该值默认为 8，该值会影响新生代内存回收的开始时间
-XX:NewRatio=3	老生代与新生代的比值，默认值为 2
-XX:MaxTenuringThreshold=8	指定年轻代对象经过 8 次 GC 后可以晋升到老年代，该值默认为 15
-XX:ParallelGCThreads=5	指定年轻代的并行收集线程数为 5 个，如果机器 CPU 核心数目小于 8，则可以设置为 CPU 核心数，如果 CPU 核心数目大于 8，则建议设置为 CPU 核心数的 1/4 左右

续表

参数	解释
-XX:ConcGCThreads=3	定义 G1 并发过程线程数为 3 个，配置更多的线程数（当然不能超过机器 CPU 核心数）会加快并发过程，但也会带来额外的同步开销。对于不同应用程序，需要在真实生产环境中通过真实的测试来确认增加线程是否能够提升 GC 性能，如果该参数未配置，JVM 会根据并行收集器中的-XX:ParallelGCThreads 参数的值来计算出默认的并行线程数（ConcGCThreads = (ParallelGCThreads + 3)/4）
-XX:G1ReservePercent=15	指定作为堆预留内存百分比为 15%，以降低目标空间溢出的风险，默认值为 10%，如果调大该值，相当于可以用的堆内存减少

接下来通过调整关键参数，观察生产环境应用 GC 的次数与时间以得出应用的最佳 JVM 参数配置，以表 9-4 所示线上集群节点为监控对象。

表 9-4 G1 GC 参数表

参数	硬件	HBase 分区服务器 JVM 配置
Slave9	32 核心 CPU，128 GB 内存	-XX:+UseG1GC -Xmx24g -Xms24g -XX:MaxDirectMemorySize=26g -XX:SurvivorRatio=6 -XX:G1ReservePercent=12 -XX:MaxGCPauseMillis=2000
Slave10	32 核心 CPU，128 GB 内存	-XX:+UseG1GC -Xmx24g -Xms24g -XX:MaxDirectMemorySize=26g -XX:SurvivorRatio=3 -XX:G1ReservePercent=12 -XX:MaxGCPauseMillis=2000
Slave11	32 核心 CPU，128 GB 内存	-XX:+UseG1GC -Xmx24g -Xms24g -XX:MaxDirectMemorySize=26g -XX:SurvivorRatio=3 -XX:G1ReservePercent=12 -XX:MaxGCPauseMillis=3000

运行一段时间后（数天），各节点 GC 时间对比如图 9-6 所示。

机器	S0	S1	E	O	P	YGC	YGCT	FGC	FGCT	GCT	运行时长（秒）
slave11	0	100	22.8	79.37	95.56	218231	14697.812	0	0	14697.812	1293745.697
slave10	0	100	7.35	74.54	96.44	240672	15122.036	0	0	15122.036	1306841.558
slave9	0	100	29.58	79.14	96.18	236676	15147.99	0	0	15147.99	1306625.867

机器	平均YGC时长（毫秒）	每小时YGC次数	每小时YGC耗时（毫秒）
slave11	67.35	608	40849
slave10	62.8	663	41659
slave9	64	652	41730

图 9-6 G1 GC 耗时对比

G1 GC 没有 Full GC，而提供了 Young GC 与 Mix GC 两种模式，Young GC 用来回收年轻代分区，Mix GC 用来回收年轻代与统计得出的收集收益高的老生代，由测试结果可得出如下结论。

- G1 GC 的收集效率远高于 CMS GC，因此建议使用 G1 GC。
- G1 的 S0、S1 两个内存区使用永远是 0 和 100，这是由于 G1 GC 只有一组逻辑上的 Survivor 区，而不像其他 GC 一样有两段明确、固定的地址空间用作 Survivor 区域，而且 Survivor 区域的分区数量是实时变动的，由测试结果也可以看到 3 台机器 GC 时间基本相同，因此 `SurvivorRatio` 参数对其影响不大。
- G1 GC 不建议使用 -Xmn 显式地指定年轻代的大小，否则会干扰 G1 收集器的暂停时间目标，同时也无法动态地分配年轻代大小。
- 配置项 `-XX:MaxGCPauseMillis` 设置得越大，则 YGC 平均耗时越长，但是 YGC 次数会降低，设置该值时应该考虑业务的平均响应时间与堆大小，如使用业务平均响应时间的 90% 作为该配置项的值，但是如果堆太大，如 32 GB，那么设置过小的值（如 `MaxGCPauseMillis=50`）明显是不合理的。

本章描述了一些常用的 HBase 调优手段，当然还有一些可以提高性能的手段，如使用堆外缓存等，很多调优都是根据生产环境现状因地制宜而实施的，所幸 HBase 集群容灾强调，支持节点的动态下线与上线，因此可以在不影响业务的情况下适时地调整配置，最终构建出一个响应迅速、运行稳定的 HBase 在线集群，在最小的开销下，发挥机器的最大性能。

第 10 章

集群间数据复制

数据的复制与备份在数据库领域一直以来都是一个非常重要的主题。在生产环境中像 MySQL 之类的数据库一般都会部署至少主备两个节点，主备的部署架构首先可以保证在硬件故障的时候数据不丢失，其次备用的从节点（slave node）可以在主节点（master node）故障时切换为主节点来保证系统的高可用性，最后还可以通过一些读写分离的策略来分担机器读写压力。

在 HBase 和 Hadoop 的生态里，由于 Hadoop 系统对数据的多备份支持以及 Hadoop 的设计思想就是运行在普通商用的机器上，因此在软件层面对硬件故障提供了天然的支持，不会与传统关系型数据库一样受单点故障的影响，也就是说，Hadoop 本身能够保证系统的高可用，单点故障不会影响 Hadoop 生态系统提供正常服务。对 Hadoop 生态来说可能需要考虑的是集群所在机房故障，一方面因为 Hadoop 集群一般不会跨机房搭建（当然，如果机房之间有速度很快的专有网络连接，就可以考虑跨机房搭建，但是网络带宽要求很高），另一方面因为 Hadoop/HBase 集群存储的数据量一般都很大，如果在在线集群上做一些数据分析，很有可能会影响在线集群的性能，因此需要搭建一个离线集群并同步线上数据来做一些统计报表等数据分析。

表 10-1 列出了 HBase 提供的常用数据复制和数据同步方式以及各自的优缺点。

表 10-1 常用的集群间数据复制以及数据同步的方式

方式	性能影响	停机	增量备份	实现难度	恢复时长
复制（replication）	很小	不需要	实时	较容易	很小
快照（snapshots）	很小	短暂（恢复时）	不支持	容易	很小
导出导入（export/import）	大	不需要	支持	容易	较长
复制表（copy table）	大	不需要	支持	容易	较长

10.1 复制

集群间复制采用数据源推的方式，一个 HBase 集群可以是一个主集群（数据生产者），也可以是一个从集群（数据消费者），或者也可以同时扮演主集群和从集群两个角色。复制是异步进行的，目的是保证数据的最终一致性。复制的粒度可以配置到列族级别，当某个分区服务器上面配置了复制的列族对应的 WAL（Write-Ahead-Log）有更新时，分区服务器会对这个 WAL 的变更（复制的单元是 WALEntry，由 WALEdit 和 WALKey 组成，WALEdit 是变更的数据，WALKey 用来记录已经复制过的集群标识符）做一系列的筛选，最后推送到所有需要复制该列族的从集群。

当数据从主集群复制到从集群的时候，复制的 WALEntry.WALKey 里面会记录下主集群的唯一标识符 clusterID，HBase 0.9.6 以及以上的版本会将所有已经消费过数据的集群 clusterID 记录下来以防止数据的循环复制（如 A→B→C→A）。

WAL 以 HLog 的形式存储在 Hadoop HDFS 中，只要任意一个从集群未复制完成，HLog 将会一直保留。每台分区服务器复制的时候会从复制队列中最旧的 WAL 开始，并且在 ZooKeeper 中记录下当前正在处理的 WAL 以及处理到的位置，每个从集群可能处理到的 WAL 与进度不一样，必须当所有从集群都已经复制完成一个 WAL 后，该 WAL 文件才能被删除。

复制的主集群与从集群规模大小通常不一样，主集群会使用随机选择复制目标分区服务器来平衡发送到每台分区服务器的数据量，如果从集群因为空间不够或者表下线等其他原因无法访问，则复制会异常或者中断，主集群会保留复制的 WAL 用于之后重试。

下面从可靠性和有序性来分析一下 HBase 的集群间复制。

- **可靠性**：因为 WAL 是存储在 Hadoop HDFS 的，HDFS 的多备份保证了数据不会丢失，HBase 复制提供至少投递一次的机制来保证每个 `WALEntry` 都能够被从集群成功收到，所以最终复制能够保证数据的可靠性。
- **有序性**：因为复制需要至少投递一次来保证数据的可靠性，所以无法保证 WAL Edit 的有序性，当一个分区服务器 A 因为异常而下线的时候，这个分区服务器上面的所有分区会转移到集群中其他分区服务器负责服务，而这个异常分区服务器 A 的未完成的复制队列也会由集群中另外一个分区服务器负责。假设分区 X 原来由分区服务器 A 负责提供服务，当分区服务器 A 异常下线后分区 X 转移到分区服务器 B 负责，如果分区服务器 A 的待复制队列也正好转移到由分区服务器 B 负责，则此时分区服务器 B 正在复制的数据可能有分区 X 上新增的数据，也可能有分区服务器 A 之前未复制完成的数据，而且分区服务器 A 未复制完成的数据可能比分区 X 上新增的数据更晚推送给从集群，因此 HBase 复制无法保证数据更新的顺序。

综上所述，某些非幂等（$f(x) = x$）的操作（如 Increments）复制到从集群后，数据可能与主集群不一致，设计应用程序的时候需要特别注意这个特性。

HBase 借鉴了 MySQL 的基于语句的复制方法，类似于 MySQL 的语句复制，整个 WALEdits（Put、Delete 等操作包含的多个单元格的数据变更）一起复制用来保证数据行的原子性。

10.1.1 集群拓扑

HBase 的复制由于其易用、低延迟、异步等特性，常用于如下场景。
- 从中心机房的主集群复制到同城或者异地机房的从集群，用来做容灾。
- 两个机房集群数据可以相互复制，互为主从，此时可以称之为主主复制，也就是说两个集群支持数据的同时写入，相互复制数据，最终保证两个集群的数据一致性，这种复制方式可以用来对在线应用做异地多活。
- 为不同业务提供的在线实时服务的多个集群可以将数据一起复制到一个离线的从集群用来做数据备份或者离线的数据分析，分析后得到的结果数据也可以复制回各业务集群以反馈服务。

多种复制策略或者拓扑可以组合使用，来满足企业各式各样的业务需求，图 10-1 描述了一个假定的复制拓扑架构，箭头表示数据的复制方向。

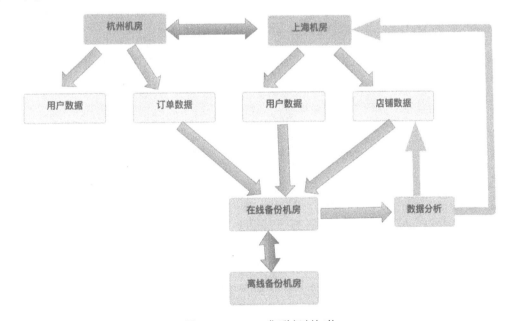

图 10-1　HBase 集群复制拓扑

10.1.2 配置集群复制

HBase 的复制配置非常简单，下面的步骤描述了开启用户行为日志表 s_behavior 的集群间数据复制的方法。

1. 修改配置文件 hbase-site.xml

代码清单 10-1 描述了需要添加到配置文件 hbase-site.xml 的配置项。

代码清单 10-1　hbase-site.xml

```xml
<property>
    <name>hbase.replication</name>
    <value>true</value>
</property>
```

2. 添加集群复制关系（peer）

在主集群 HBase shell 运行如代码清单 10-2 所示命令即可添加集群复制关系，数据即可从主集群流向配置的从集群，格式为：

```
add_peer [peerID] [hbase.zookeeper.quorum:hbase.zookeeper.property.clientPort:zookeeper.znode.parent]
```

第一个参数为一个自定义数字，用来唯一标识这个复制关系。第二个参数全部为与 ZooKeeper 相关配置项，均可以在 hbase-site.xml 配置文件找到。

代码清单 10-2　添加复制关系

```
add_peer '1','master1,master2,slave1:2181:/hbase'
```

3. 启用 HBase 表或者列族复制

代码清单 10-3 描述了如何对用户行为日志表的列族 pc 开启复制。注意，需要在主集群和从集群都执行开启列族复制脚本。

代码清单 10-3　开启列族复制

```
disable 's_behavior'
alter 's_behavior' , {NAME=>"pc", REPLICATION_SCOPE=>"1", KEEP_DELETED_CELLS => 'TRUE'}
enable 's_behavior'
```

此外，还可以在表创建的时候指定开启复制：

```
create 's_test', {NAME => 'cf', REPLICATION_SCOPE => '1', KEEP_DELETED_CELLS => 'TRUE'}
```

其中 REPLICATION_SCOPE 是一个列族级别的配置，可以取值 0、1 或 2，具体含义如下。

- 0：表示不开启列族的复制。
- 1：表示开启列族的复制，但是不保证 WAL 到达从集群的顺序。
- 2：表示开启列族的有序复制，这是 HBase 1.5 引入的一个新功能，这个功能让 WAL 能够以到达主集群的顺序到达从集群，这就适合一些对有序性有要求的应用，但是当然也会有一定的开销，对复制性能有一定影响。

配置成功后在 HBase shell 控制台输入命令 `status 'replication'` 即可查看复制状态，输出结果如代码清单 10-4 所示。

代码清单 10-4　查看集群复制状态

```
hbase(main):010:0> status 'replication'
version 1.2.4
3 live servers
   master1:
      SOURCE: PeerID=1, AgeOfLastShippedOp=0, SizeOfLogQueue=0,
TimeStampsOfLastShippedOp=Mon Dec 25 11:28:14 CST 2017, Replication Lag=0
      SINK: AgeOfLastAppliedOp=0, TimeStampsOfLastAppliedOp=Mon Dec 25 11:10:11 CST 2017
   master2:
      SOURCE: PeerID=1, AgeOfLastShippedOp=0, SizeOfLogQueue=0,
TimeStampsOfLastShippedOp=Mon Dec 25 11:28:14 CST 2017, Replication Lag=0
      SINK: AgeOfLastAppliedOp=0, TimeStampsOfLastAppliedOp=Mon Dec 25 11:10:10 CST 2017
   slave1:
      SOURCE: PeerID=1, AgeOfLastShippedOp=0, SizeOfLogQueue=0,
TimeStampsOfLastShippedOp=Mon Dec 25 11:28:14 CST 2017, Replication Lag=0
      SINK: AgeOfLastAppliedOp=0, TimeStampsOfLastAppliedOp=Mon Dec 25 11:10:14 CST 2017
```

如果配置不成功或者从集群不在线等，则查看状态输出如代码清单 10-5 所示，可以看到上次复制时间 `TimeStampsOfLastShippedOp` 为 `Thu Jan 01 08:00:00 CST 1970`，对应的时间戳为 0，表示从未复制过。

代码清单 10-5　集群复制状态异常

```
hbase(main):002:0> status 'replication'
version 1.2.4
3 live servers
   master1:
      SOURCE: PeerID=1, AgeOfLastShippedOp=0, SizeOfLogQueue=1,
TimeStampsOfLastShippedOp=Thu Jan 01 08:00:00 CST 1970, Replication Lag=1514171477661
      SINK : AgeOfLastAppliedOp=0, TimeStampsOfLastAppliedOp=Mon Dec 25 11:10:11 CST 2017
   master2:
      SOURCE: PeerID=1, AgeOfLastShippedOp=0, SizeOfLogQueue=253,
TimeStampsOfLastShippedOp=Thu Jan 01 08:00:00 CST 1970, Replication Lag=1514171476826
      SINK: AgeOfLastAppliedOp=0, TimeStampsOfLastAppliedOp=Mon Dec 25 11:10:10 CST 2017
   slave1:
      SOURCE: PeerID=1, AgeOfLastShippedOp=0, SizeOfLogQueue=1,
TimeStampsOfLastShippedOp=Thu Jan 01 08:00:00 CST 1970, Replication Lag=1514171477419
      SINK: AgeOfLastAppliedOp=0, TimeStampsOfLastAppliedOp=Mon Dec 25 11:10:14 CST 2017
```

4. 复制相关命令

在 HBase shell 输入 `help 'replication'` 命令即可查看 HBase shell 支持的复制命令。下面列出了常用的复制相关命令。

- `add_peer <ID> < CLUSTER_KEY >`：其中 ID 表示该复制关系的唯一标识符，一般使用短整型数字；CLUSTER_KEY 由如下 3 个配置项组成，中间用英文冒号分隔，这 3 个配置项全部可以在 hbase-site.xml 中找到：

  hbase.zookeeper.quorum:hbase.zookeeper.property.clientPort: zookeeper.znode.parent

  例如，`add_peer '1', 'master1,master2,slave1:2181:/hbase'`，hbase.zookeeper.quorum 值为 master1,master2,slave1，hbase.zookeeper.property.clientPort 值为 2181，zookeeper.znode.parent 值为 hbase，该配置项也是 HBase 在 ZooKeeper znode 根节点。

- `list_peers`：该命令会列出在当前集群中所有的复制关系。

- `disable_peer <ID>`：停用 ID 代表的复制关系，停用后，HBase 将不会再推送 WAL 变更到目的集群，但是仍然会记录下需要复制的 WAL，以防止复制关系再次被启用，这些 WAL 会被保留直到这个停用的复制关系不存在了（即用 remove_peer 移除了）。例如，`disable_peer '1'`。

- `enable_peer <ID>`：与 disable_peer 相反，启用 ID 代表的已经停用的复制关系。例如，`enable_peer '1'`。

- `remove_peer <ID>`：停用并且移除复制关系，HBase 将不会再推送 WAL 变更到目的集群，并且 HBase 也不会保留之前需要复制的 WAL。例如，`remove_peer '1'`。

- `enable_table_replication <TABLE_NAME>`：启用表所有列族的复制，如果从集群没有该表，则 HBase 会基于主集群表模式创建相同的表。例如，`enable_table_replication 's_behavior'`。

- `disable_table_replication <TABLE_NAME>`：停用表所有列族的复制。例如，`disable_table_replication 's_behavior'`。

- `list_replicated_tables`：列出所有复制的表。

- `set_peer_tableCFs <ID>, <表:列族>`：add_peer 添加的复制连接默认复制集群所有的表，如果只想复制某个表或者表的某个列族，则可以使用该命令。例如，`set_peer_tableCFs '1','s_behavior:ph'`。

- `remove_peer_tableCFs <ID>, <表:列族>`：与 set_peer_tableCFs 相对，移除某个表列族在 ID 代表的复制关系的复制。例如，`remove_peer_tableCFs '1', 's_behavior:ph'`。

10.1.3 验证复制数据

数据最终的完整性与一致性无疑是复制首先需要保证的。对于如何验证数据的完整性与一致性，HBase 提供了一个名为 VerifyReplication 的 MapReduce 作业，可以用来对主从集群中复制的数据做一个系统的比较，命令格式如下：

```
${HBASE_HOME}/bin/hbase org.apache.hadoop.hbase.mapreduce.replication.VerifyReplication
--starttime=<timestamp> --stoptime=<timestamp> <ID> <tableName>
```

例如：

```
hbase org.apache.hadoop.hbase.mapreduce.replication.VerifyReplication --starttime=
1265875194289 --stoptime=1510504750672 1 s_behavior
```

该 MapReduce 任务执行结果如代码清单 10-6 所示，输出结果的倒数第 5 行 GOODROWS=3 表示主集群与从集群表 s_behavior 数据一致的行数，同样 BADROWS 表示主从集群表数据不一致的行数。

代码清单 10-6　集群复制验证结果

```
2017-11-12 20:36:19,998 INFO  [main] mapreduce.JobSubmitter: number of splits:1
2017-11-12 20:36:20,198 INFO  [main] mapreduce.JobSubmitter: Submitting tokens
for job: job_1510490151985_0001
2017-11-12 20:36:20,480 INFO  [main] impl.YarnClientImpl: Submitted application
application_1510490151985_0001
2017-11-12 20:36:20,515 INFO  [main] mapreduce.Job: The url to track the job:
http://wxmaster2:8088/proxy/application_1510490151985_0001/
2017-11-12 20:36:20,516 INFO  [main] mapreduce.Job: Running job: job_15104901519
85_0001
2017-11-12 20:36:27,601 INFO  [main] mapreduce.Job: Job job_1510490151985_0001
running in uber mode : false
2017-11-12 20:36:27,602 INFO  [main] mapreduce.Job:  map 0% reduce 0%
2017-11-12 20:36:34,659 INFO  [main] mapreduce.Job:  map 100% reduce 0%
2017-11-12 20:36:34,667 INFO  [main] mapreduce.Job: Job job_1510490151985_0001
completed successfully
2017-11-12 20:36:34,758 INFO  [main] mapreduce.Job: Counters: 44
        File System Counters
                FILE: Number of bytes read=0
                FILE: Number of bytes written=144441
                FILE: Number of read operations=0
                FILE: Number of large read operations=0
                FILE: Number of write operations=0
                HDFS: Number of bytes read=67
                HDFS: Number of bytes written=0
                HDFS: Number of read operations=1
                HDFS: Number of large read operations=0
```

```
                HDFS: Number of write operations=0
        Job Counters
                Launched map tasks=1
                Rack-local map tasks=1
                Total time spent by all maps in occupied slots (ms)=14088
                Total time spent by all reduces in occupied slots (ms)=0
                Total time spent by all map tasks (ms)=3522
                Total vcore-seconds taken by all map tasks=3522
                Total megabyte-seconds taken by all map tasks=14426112
        Map-Reduce Framework
                Map input records=3
                Map output records=0
                Input split bytes=67
                Spilled Records=0
                Failed Shuffles=0
                Merged Map outputs=0
                GC time elapsed (ms)=33
                CPU time spent (ms)=1210
                Physical memory (bytes) snapshot=190156800
                Virtual memory (bytes) snapshot=4078424064
                Total committed heap usage (bytes)=317718528
        HBase Counters
                BYTES_IN_REMOTE_RESULTS=120
                BYTES_IN_RESULTS=120
                MILLIS_BETWEEN_NEXTS=281
                NOT_SERVING_REGION_EXCEPTION=0
                NUM_SCANNER_RESTARTS=0
                NUM_SCAN_RESULTS_STALE=0
                REGIONS_SCANNED=1
                REMOTE_RPC_CALLS=3
                REMOTE_RPC_RETRIES=0
                ROWS_FILTERED=0
                ROWS_SCANNED=3
                RPC_CALLS=3
                RPC_RETRIES=0
        org.apache.hadoop.hbase.mapreduce.replication.VerifyReplication$Verifier$Counters
                GOODROWS=3
        File Input Format Counters
                Bytes Read=0
        File Output Format Counters
                Bytes Written=0
```

> **注意** 运行 VerifyReplication 之前需要执行如下命令先启动 Hadoop ResourceManager 和 NodeManager：
>
> ```
> /home/hadoop/hadoop-2.6.5/sbin/yarn-daemon.sh start resourcemanager
> /home/hadoop/hadoop-2.6.5/sbin/yarn-daemon.sh start nodemanager
> ```

10.1.4 复制详解

HBase 复制由于其易用、低延迟等特性，在 HBase 的数据备份与同步中占有了一个重要的位置。本节分析 HBase 复制源码，感兴趣的读者可以继续往下阅读，初学者建议跳过本节，待对 HBase 有一定了解后可回头参照 HBase 源码一起阅读。

1. 复制初始化

当使用 Shell 脚本启动 HBase HRegionServer 进程的时候，HBase 复制相关类就开始初始化了，初始化完成后，每个从集群复制连接都对应一个 ReplicationSource 线程，ReplicationSource 内部又会为每个 WAL Group 创建一个后台线程 ReplicationSourceWorkThread，该后台线程一直扫描对应的 WAL，只要 WAL 有数据变更（如 HBase Client 调用 Put、Delete 等操作写入数据）就会读取 WAL 内容进行复制，图 10-2 描述了这一初始化过程。

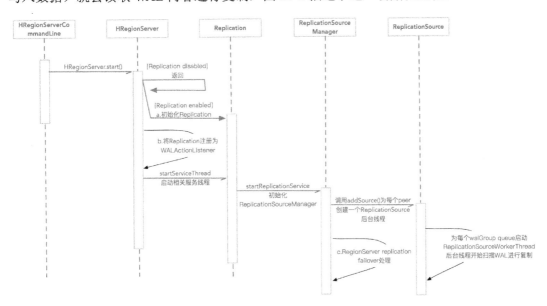

图 10-2 HBase 复制初始化过程

（1）初始化复制。
- 新建与初始化类 ReplicationQueues replicationQueues，实现类为 ReplicationQueuesZKImpl，该类用于控制分区服务器需要复制的 WAL 队列，在 ZooKeeper 中会创建一个与分区服务器同名的 znode，用来保存这个需要复制的 WAL 文件队列。
- 新建与初始化类 ReplicationPeers replicationPeers，实现类为 Replication-

`PeersZKImpl`，用于管理复制的状态，在 ZooKeeper 中会创建一个 `/hbase/replication/peers/1` 这样的 znode 用来保存这个从集群连接（peer）复制的状态（`ENABLED` 或 `DISABLED`）以及需要复制的表和列族。

- 新建类 `ReplicationTracker replicationTracker`，实现类为 `ReplicationTrackerZKImpl`，该类监听 ZooKeeper 节点变化导致复制状态变更的事件，如某台分区服务器宕机需要容灾。
- 新建类 `ReplicationSourceManager replicationManager`，用来管理所有的 `ReplicationSource`。

（2）将 `Replication` 类注册为 `WALActionListener`。这样，当 WAL 有回滚（roll）的时候，会转调到 `ReplicationSource` 将 WAL 日志添加到 `walGroup` 队列 `Map`，以便 `ReplicationSourceWorkThread` 后台扫描并发送 WAL 内容到从集群。

（3）分区服务器复制容灾处理。检查是否有 `replicationQueues` 所属的分区服务器已经宕机，如果有，则当前分区服务器接管宕机的分区服务器的复制。

2. WAL 复制

WAL 复制是 HBase 复制的核心步骤，数据变更将在这一步从主集群复制到从集群，图 10-3 描述了这一复制过程。

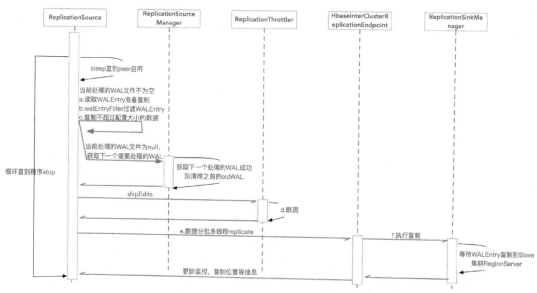

图 10-3　HBase WAL 复制

（1）读取 `WALEntry` 准备复制。
- WAL 文件复制读取进度保存在 ZooKeeper。
- `WALEntry` 由 `WALEdit` 和 `WALKey` 组成。

- WALEdit 是真实需要复制的数据。
- WALKey 记下了复制过的从集群的唯一标示符 clusterId 防止循环复制。

（2）walEntryFilter 过滤 WALEntry。

- SystemTableWALEntryFilter 过滤系统表的 WALEntry。
- ScopeWALEntryFilter 过滤主集群列族 REPLICATION_SCOPE!=1 的键值对。
- TableCfWALEntryFilter 过滤从集群没有开启复制的列族下的键值对。

（3）复制不超过配置大小的数据。

- 每次读取的数据量大小不超过 replication.source.size.capacity 配置的值（默认 64 MB）。
- 每次读取的 WALEntry 数量不超过 replication.source.nb.capacity 配置的值（默认 25000）。

（4）限流。

- 如果限流周期(100 ms)内复制了超过带宽限制的数据量，则休眠，复制数据量/带宽×100 ms。
- 如果限流周期内复制的数据加上该次需要复制的数据量大于限流带宽，则休眠到下一个周期再复制。

（5）数据分批多线程复制。

- 分批数为 A（hbase.replication.source.maxthreads 配置的值，默认为 10）、B（WALEntry/100+1）和 C（slave 集群分区服务器数×replication.source.ratio 配置的比例）三者中的最小值。
- 按 RegionName 分批，每批启动一个 Replicator（Callable 线程）提交到线程池开始执行复制。

（6）执行复制。

- 随机获取从集群的一台分区服务器执行复制。
- 调用 ReplicationProtbufUtil.replicateWALEntry 发送 RPC 请求执行复制。

3. 分区服务器复制容灾

分区服务器由于硬件的故障或者负载过高而宕机是很常见的，HBase 复制使用 ZooKeeper 来保证高可用。当一台分区服务器故障离线时，ZooKeeper 会负责协调管理复制队列的转移。

（1）以一个主集群包含 3 台分区服务器复制到一个 peer_id=1 的从集群为例，图 10-4 描述了正常情况下某个时间点上 ZooKeeper 的 znode 层次结构，所有的分区服务器的 znode 都包含一个 peer_id=1 的 znode，这个 znode 包含一系列需要复制的 WAL 文件 HDFS 路径（格式是"主机,端口,时间戳.xxx"，如 master1,16010,1514193806746.default.1514193808901。注意","会被编码成"%2C"）。

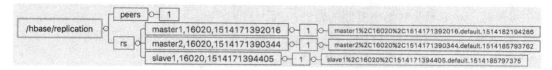

图 10-4　HBase 正常复制 ZooKeeper znode

图 10-5 描述了分区服务器 master1 包含多个 WAL 日志的情况。

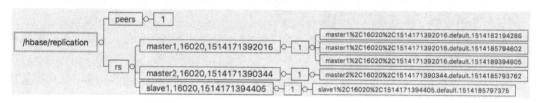

图 10-5　HBase 正常复制 master1 多个 WAL

（2）主集群上的每台分区服务器都会监控集群中其他分区服务器是否宕机，一旦有分区服务器宕机，其他分区服务器都会收到通知，然后这些分区服务器就会竞争去 ZooKeeper 上创建一个名为 `lock` 的 znode，竞争成功的分区服务器会负责这个离线分区服务器剩余 WAL 的复制权，如图 10-6 所示，假设分区服务器 master1 离线，剩下的 2 台分区服务器会竞争创建 znode，假设分区服务器 master2 竞争成功，则 master2 会将 master1 的复制队列迁移到它自己的复制队列下面。

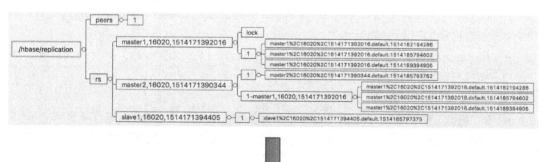

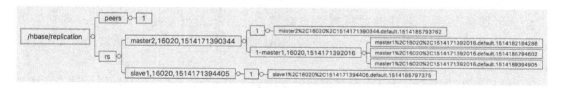

图 10-6　分区服务器 master1 宕机

（3）接下来分区服务器 master2 会为复制过来的 master1 的 WAL 复制队列（queue）创建一个 `ReplicationSourceWorkerThread` 后台线程复制这些 WAL，复制的流程同样如前所述（读取 WALEntry，过滤 WALEntry，执行 WALEntry 复制）。当最后一个 WAL 复制完成后，这个复制队列的 znode 会被删除，因为这个复制队列所属的分区服务器已经离线，所以肯定不会有 WAL 再添加到这个复制队列。

（4）假设接下来分区服务器 master2 也离线了，那么分区服务器 slave1 也会执行同样的接管过程，最终 ZooKeeper znode 树形结构如图 10-7 所示。

图 10-7　分区服务器 master2 宕机

（5）即使之后分区服务器 master1 和 master2 恢复服务，已经被接管的 WAL 复制队列仍然会由 slave1 负责，如图 10-8 所示。

图 10-8　分区服务器 master1 和 master2 恢复

10.2　快照

除了实时备份之外，定期备份归档也是数据库系统保证数据可靠性或者容灾恢复的重要手段。例如，生成环境中程序出现 bug，需要用最近的一个备份来恢复数据，一般 DBA 都会对数据库做定期备份，如一周一次全量备份、一天一次增量备份，快照（snapshot）就提供了一个很好的全量备份的方式。

一个快照是系统在某一时刻的数据的一个完整镜像。快照最简单的实现方式就是对系统加写锁，然后复制出一个数据的备份，这样这个数据的备份就会是某一时刻完全一致性的数据。因为需要加写锁，所以制作镜像的过程系统只能提供数据读操作，等数据复制完成之后才能释放写锁。当数据量大时，数据复制备份必定会花费大量时间，对系统性能影响巨大，这种实现方式对于在线实时系统是无法容忍的。

HBase 快照可以让你对 HBase 表做一个镜像，在常数时间内完成，并且对分区服务器几乎没什么影响，快照记录了该时刻所有的数据，快照的克隆以及恢复操作不会涉及数据的复制，同样，将快照导出到另外一个集群也不会影响到分区服务器。

快照之所以能在常数时间内完成，是因为它只是一组元数据（metadata）的集合。这些元数据描述了快照制作时表所用到的 HFile 的文件名，因为 HFile 生成后是不会修改的，只有在进行主压缩（compact）或者拆分（split）的时候才可能会对 HFile 进行删除操作，所以快照制作完成后只需要提醒系统在 compact 或者 split 的时候不删除这些 HFile 即可。

10.2.1 配置快照

启用 HBase 快照配置非常简单，只需在 `hbase-site.xml` 中添加代码清单 10-7 所示的配置项即可。

代码清单 10-7 配置集群支持快照

```xml
<property>
    <name>hbase.snapshot.enabled</name>
    <value>true</value>
</property>
```

10.2.2 管理快照

HBase 提供了 Java API 和 HBase Shell 两种方式来管理快照，使用简单。本节介绍如何使用 HBase Shell 来管理快照。

1. 制作快照

无论 HBase 表是否在线，我们都可以制作表的快照。注意，HBase 的新写入的数据先写入 MemStore，可能还未刷新到磁盘落地到 HFile，因此制作快照的时候也需要把 MemStore 里面的这部分数据先刷新到 HFile，也就是说需要将 MemStore 刷新成 HFile（这里就涉及需要将 MemStore 加锁禁止写入），然后将 HFile 路径记录下来。如果能够容忍 MemStore 里面这部分数据被排除在这个快照之外，则可以使用 `SKIP_FLUSH` 参数来省略刷新 MemStore 到 HFile 文件的操作。代码清单 10-8 分别演示了使用与不使用 `SKIP_FLUSH` 对用户行为数据表 `s_behavior` 制作快照。

代码清单 10-8 制作快照

```
hbase(main):026:0* snapshot 's_behavior', 's_behaviorSnapshot-201801281223'
0 row(s) in 0.3500 seconds

hbase(main):028:0> snapshot 's_behavior', 's_behaviorSnapshot-201801281224',
{SKIP_FLUSH => true}
0 row(s) in 0.3150 seconds
```

2. 查看快照

与其他 HBase Shell 命令（如 list 和 list_peer）类似，list_snapshots 可以查看 HBase 集群所有已经创建的快照，代码清单 10-9 列出了前面创建的两个快照。

代码清单 10-9　查看快照

```
hbase(main):030:0> list_snapshots
SNAPSHOT                              TABLE + CREATION TIME
 s_behaviorSnapshot-201801281223       s_behavior (Sun Jan 28 12:15:40 +0800 2018)
 s_behaviorSnapshot-201801281224       s_behavior (Sun Jan 28 12:16:44 +0800 2018)

2 row(s) in 0.0330 seconds

=> ["s_behaviorSnapshot-201801281223", "s_behaviorSnapshot-201801281224"]
```

3. 删除快照

删除快照同样不涉及 HFile 数据文件的操作，因此同样在常数时间可以完成，代码清单 10-10 删除了名为 s_behaviorSnapshot-201801281224 的快照。

代码清单 10-10　删除快照

```
hbase(main):031:0> delete_snapshot "s_behaviorSnapshot-201801281224"
0 row(s) in 0.0310 seconds
```

4. 从快照克隆表

快照创建后可以创建一个新表，新表的数据与快照镜像时刻的数据一模一样，并且新表的数据变更不会影响到快照以及旧表，如代码清单 10-11 所示，用快照 s_behaviorSnapshot-201801281223 克隆出新表 s_behavior_clone。

代码清单 10-11　克隆表

```
hbase(main):033:0> clone_snapshot 's_behaviorSnapshot-201801281223','s_behavior_clone'
0 row(s) in 0.3900 seconds
```

5. 从快照恢复数据

恢复操作需要将 HBase 表下线，恢复后表的状态会回滚到快照创建时刻，包括数据和表模式（schema）都会被回滚，如代码清单 10-12 所示。

代码清单 10-12　快照恢复

```
hbase(main):035:0> disable 's_behavior'
0 row(s) in 2.2770 seconds

hbase(main):036:0> restore_snapshot 's_behaviorSnapshot-201801281223'
0 row(s) in 0.2710 seconds
```

6. 将快照导出到其他集群

HBase 提供了一个 ExportSnapshot 工具类用来将快照复制到另外一个集群。复制的数据包括 HFile、HLog 以及快照元数据。类似于 distcp，这个工具类是一个基于文件复制的 MapReduce 作业（执行之前需要先启动 Hadoop ResourceManager 和 NodeManager），因此 HBase 集群无须下线，但是当数据量较大时，需要消耗一定的资源。

代码清单 10-13 使用 16 个线程将快照 s_behaviorSnapshot-201801281223 导出到 HBase 集群 cluster2。

代码清单 10-13　快照导出

```
hbase org.apache.hadoop.hbase.snapshot.ExportSnapshot -snapshot s_behaviorSnapshot
-201801281223 -copy-to hdfs://cluster2:8082/hbase -mappers 16
```

导出快照数据涉及数据传输，可能占用大量带宽，该工具类可接收参数 -bandwidth 以限定导出命令消耗的最大带宽（单位 MB/s），如代码清单 10-14 所示。

代码清单 10-14　快照导出限制带宽

```
hbase org.apache.hadoop.hbase.snapshot.ExportSnapshot -snapshot s_behaviorSnapshot
-201801281223 -copy-to hdfs://cluster2:8082/hbase -mappers 16 -bandwidth 200
```

10.3　导出和导入

HBase 提供了 Export MapReduce 作业用来把 HBase 的表导出（export）为文件，然后使用 Import MapReduce 作业来把文件导入（import）同一个或者另外一个集群的 HBase 表中。导出和导入涉及数据的读取、传输，因此对性能影响相对较大，一般在离线集群使用，因为导入导出需要使用到 MapReduce 作业，所以需要在 Hadoop 集群使用如下命令启动 ResourceManager 和 NodeManager：

```
/home/hadoop/hadoop-2.6.5/sbin/yarn-daemon.sh start resourcemanager
/home/hadoop/hadoop-2.6.5/sbin/yarn-daemon.sh start nodemanager
```

10.3.1　导出

HBase Export 命令参数较多，先看下代码清单 10-15 所示的使用说明。

代码清单 10-15　导出命令说明

```
[hadoop@master1 root]$ hbase org.apache.hadoop.hbase.mapreduce.Export
ERROR: Wrong number of arguments: 0
Usage: Export [-D <property=value>]* <tablename> <outputdir> [<versions> [<starttime>
[<endtime>]] [^[regex pattern] or [Prefix] to filter]]
```

```
Note: -D properties will be applied to the conf used.
For example:
  -D mapreduce.output.fileoutputformat.compress=true
  -D mapreduce.output.fileoutputformat.compress.codec=org.apache.hadoop.io.compress.GzipCodec
  -D mapreduce.output.fileoutputformat.compress.type=BLOCK
  Additionally, the following SCAN properties can be specified
to control/limit what is exported..
  -D hbase.mapreduce.scan.column.family=<familyName>
  -D hbase.mapreduce.include.deleted.rows=true
  -D hbase.mapreduce.scan.row.start=<ROWSTART>
  -D hbase.mapreduce.scan.row.stop=<ROWSTOP>
For performance consider the following properties:
  -Dhbase.client.scanner.caching=100
  -Dmapreduce.map.speculative=false
  -Dmapreduce.reduce.speculative=false
For tables with very wide rows consider setting the batch size as below:
  -Dhbase.export.scanner.batch=10
```

1. 导出整个表

（1）导出到 HBase 所在 HDFS。默认情况下导出的文件会保存到 HBase 数据存储的 Hadoop 集群的 HDFS 目录，如下代码将表 s_behavior 导出到 HDFS 目录 /home/hadoop/s_behavior：

```
hbase org.apache.hadoop.hbase.mapreduce.Export 's_behavior' /home/hadoop/s_behavior
```

使用如下 Hadoop fs 命令可以查看导出的文件：

```
[hadoop@master1 ~]$ hadoop fs -ls /home/hadoop
Found 1 items
drwxr-xr-x   - hadoop supergroup          0 2017-11-13 19:23 /home/hadoop/s_behavior
```

（2）导出到指定 HDFS。使用如下命令可以将文件导出到目标 HDFS 集群，以方便导入：

```
hbase org.apache.hadoop.hbase.mapreduce.Export 's_behavior' hdfs://master1:9000/home/hadoop/s_behavior
```

（3）导出到本地文件。使用如下命令可以将文件导出到本地文件，以方便复制备份：

```
hbase org.apache.hadoop.hbase.mapreduce.Export 's_behavior' file:///home/hadoop/s_behavior
```

2. 按时间区间导出

如果只需要导出表在某个时间范围内的数据以用作离线分析，则可以指定导出数据的开始和结束时间。如下命令导出 s_behavior 表数据版本为 1，时间区间在 0<=时间戳<1505959355000 的数据：

```
hbase org.apache.hadoop.hbase.mapreduce.Export 's_behavior' /home/hadoop/s_behavior_
time 1 0 1505959355000
```

10.3.2 导入

导入命令相对简单，代码清单 10-16 列出了 Import 命令使用说明。

代码清单 10-16　导入命令说明

```
[hadoop@wxmaster1 root]$ hbase org.apache.hadoop.hbase.mapreduce.Import
ERROR: Wrong number of arguments: 0
Usage: Import [options] <tablename> <inputdir>
By default Import will load data directly into HBase. To instead generate
HFiles of data to prepare for a bulk data load, pass the option:
  -Dimport.bulk.output=/path/for/output
 To apply a generic org.apache.hadoop.hbase.filter.Filter to the input, use
  -Dimport.filter.class=<name of filter class>
  -Dimport.filter.args=<comma separated list of args for filter
 NOTE: The filter will be applied BEFORE doing key renames via the HBASE_IMPORTE
R_RENAME_CFS property. Futher, filters will only use the Filter#filterRowKey(byt
e[] buffer, int offset, int length) method to identify  whether the current row
needs to be ignored completely for processing and  Filter#filterKeyValue(KeyValue)
method to determine if the KeyValue should be added; Filter.ReturnCode#INCLUDE
E and #INCLUDE_AND_NEXT_COL will be considered as including the KeyValue.
To import data exported from HBase 0.94, use
  -Dhbase.import.version=0.94
For performance consider the following options:
  -Dmapreduce.map.speculative=false
  -Dmapreduce.reduce.speculative=false
  -Dimport.wal.durability=<Used while writing data to hbase. Allowed values are
the supported durability values like SKIP_WAL/ASYNC_WAL/SYNC_WAL/...>
```

如下两行命令分别将 HBase 集群所在的 HDFS 文件/home/hadoop/s_behavior 和本地文件/home/hadoop/s_behavior_local 导入到表 s_behavior_import。注意导入之前需要确保表 s_behavior_import 已经存在：

```
hbase org.apache.hadoop.hbase.mapreduce.Import 's_behavior_import' /home/hadoop/s_behavior
hbase org.apache.hadoop.hbase.mapreduce.Import 's_behavior_import'
file:///home/hadoop/s_behavior_local
```

10.4　复制表

HBase CopyTable 支持同一个集群中不同表之间数据的复制，也可以将一个集群中表的数据复制到另一个集群的表，复制之前需要先创建表，复制表也是一个 MapReduce 任务，因此执行之前需要先启动 Hadoop ResourceManager 和 NodeManager。

代码清单10-17列出了CopyTable命令使用说明。

代码清单10-17 复制表命令说明

```
[hadoop@wxmaster1 root]$ hbase org.apache.hadoop.hbase.mapreduce.CopyTable
Usage: CopyTable [general options] [--starttime=X] [--endtime=Y] [--new.name=NEW]
[--peer.adr=ADR] <tablename>

Options:
 rs.class     hbase.regionserver.class of the peer cluster
              specify if different from current cluster
 rs.impl      hbase.regionserver.impl of the peer cluster
 startrow     the start row
 stoprow      the stop row
 starttime    beginning of the time range (unixtime in millis)
              without endtime means from starttime to forever
 endtime      end of the time range.  Ignored if no starttime specified.
 versions     number of cell versions to copy
 new.name     new table's name
 peer.adr     Address of the peer cluster given in the format
              hbase.zookeeer.quorum:hbase.zookeeper.client.port:zookeeper.znode.parent
 families     comma-separated list of families to copy
              To copy from cf1 to cf2, give sourceCfName:destCfName.
              To keep the same name, just give "cfName"
 all.cells    also copy delete markers and deleted cells
 bulkload     Write input into HFiles and bulk load to the destination table

Args:
 tablename    Name of the table to copy

Examples:
 To copy 'TestTable' to a cluster that uses replication for a 1 hour window:
 $ bin/hbase org.apache.hadoop.hbase.mapreduce.CopyTable --starttime=1265875194289
--endtime=1265878794289 --peer.adr=server1,server2,server3:2181:/hbase -families
=myOldCf:myNewCf,cf2,cf3 TestTable
For performance consider the following general option:
  It is recommended that you set the following to >=100. A higher value uses more
memory but
  decreases the round trip time to the server and may increase performance.
    -Dhbase.client.scanner.caching=100
  The following should always be set to false, to prevent writing data twice,
which may produce
  inaccurate results.
    -Dmapreduce.map.speculative=false
```

（1）同集群复制表到新表。下面的命令将s_behavior表复制到新表s_behavior_copy：

```
hbase org.apache.hadoop.hbase.mapreduce.CopyTable --new.name=s_behavior_copy
s_behavior
```

（2）复制到其他集群。下面的命令将 s_behavior 表复制到集群 ZooKeeper 地址为 omaster1,omaster2,oslave1:2181 的新表 s_behavior_copy：

```
hbase org.apache.hadoop.hbase.mapreduce.CopyTable --peer.adr=omaster1,omaster2,
oslave1:2181:/hbase --new.name=s_behavior_copy  s_behavior
```

（3）按时间区间复制。下面的命令将 s_behavior 表满足条件（1510484706537<=时间戳<1510484750672）的数据复制到集群 ZooKeeper 地址为 omaster1,omaster2,oslave1:2181 的新表 s_behavior_copy：

```
hbase org.apache.hadoop.hbase.mapreduce.CopyTable --starttime=1510484706537
--endtime=1510484750672 --peer.adr=omaster1,omaster2,oslave1:2181:/hbase
--new.name=s_behavior_copy  s_behavior
```

第 11 章

监控

监控是一个成熟的软件系统不可或缺的一个组件,通过监控系统可以了解系统的负载、预警、解决异常问题、采集系统核心指标、调优与合理利用资源,最终实现系统的稳定和可靠。

监控可以分为机器监控与业务监控。

- 机器监控:如 CPU、网络、内存、磁盘等机器的核心指标。
- 业务监控:主要是与业务相关的监控,如 HBase 的分区服务器进程、业务应用是否正常提供服务、业务负载是否均衡以及业务异常出现的上下文等。

一般大公司内部都有自己的监控系统,如阿里的 Alimonitor 可以同时做到机器监控与业务监控并且集成了预警,小米针对 HBase 的统一运维监控解决方案 Minos(已经开源)。市面上也有很多开源的监控软件可以使用,如 Ganglia、Zabbix 等。

Hadoop 与 HBase 集群少则数十节点,多则上千节点,而且每个节点上可以运行多个不同的进程。如何将这些资源管理起来以及出现问题时如何及时预警是一个监控系统必须解决的问题。本章主要通过 Hadoop 与 HBase 自带的 Http 监控页面与接口介绍业务监控相关指标。

11.1 Hadoop 监控

Hadoop 提供了多种监控方式,如 Web 页面、JMX、客户端 API 等,不同的监控集群少则数十节点,多则上千节点,而且每个节点上可以运行多个不同的进程,机器越多,发生问题的频率和种类就越多,单纯靠人力去运维、监控这些机器对运维人员会是一个灾难。因此,做到平常无须运维,出现问题及时预警是监控系统的目标。

11.1.1 Web 监控页面

Hadoop 自带的 Web 页面提供了最基本的集群监控信息，包括节点运行状态、节点磁盘空间使用率、节点 JVM 内存使用情况、MapReduce 任务运行情况等，Web 监控页面地址为 http://master1:50070。

如图 11-1 至图 11-2 所示，Web 监控页面包含了概览、节点列表、镜像、启动进度和工具 5 个标签页，各标签页内容如下。

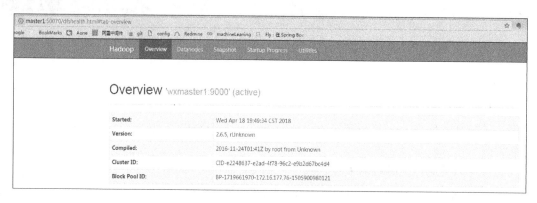

图 11-1 Hadoop 监控——概览

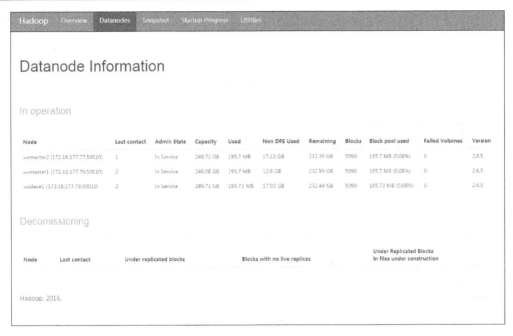

图 11-2　Hadoop 监控——数据节点

- 概览：集群启动信息、Hadoop 版本、HDFS 已使用空间、可用空间、目录与文件总数等。
- 节点列表：集群中所有节点相关的信息，如上次心跳时间、容量使用情况、分块数、故障磁盘数目以及正在下线的节点列表等。
- 镜像：Hadoop HDFS 镜像列表。
- 启动进度：Hadoop NameNode 启动加载的相关信息。
- 工具：可以用来浏览 HDFS 存储的文件，查看当前节点 Hadoop 相关运行日志。

11.1.2　JMX 监控

由于 Web 监控页面只能由应用管理员或者运维人员手动查看以确定系统是否存在异常，因此 Hadoop 还提供了一个 JMX 监控接口，用来查看包括集群节点信息、内存状态、JVM 堆内存使用情况，监控地址为：

```
NameNode: http://namenodehost:50070/jmx
DataNode: http://datanodehost:50075/jmx
```

开发人员可以将这个 JMX 接口与企业自身的监控系统打通，通过定时请求来判断 Hadoop 集群是否存在异常。这两个 URL 地址返回了很多的 JMX 相关信息，其中很多信息可能是监控不感兴趣的或者不需要监控的。如果仅需要某些监控指标数据，则该 JMX 接口提供了 3 个查询参数，即 `callback`、`get` 和 `qry`。

（1）`callback`：Web 开发中经常会使用 JSONP 来解决浏览器的数据跨域访问限制，`callback` 请求可以使得 Hadoop JMX 监控接口返回的 JSON 数据作为参数值传递给 `callback` 参数所表示的 JavaScript 函数，例如 `https://master1:50070/jmx?callback=hbasepractice` 请求返回数据如代码清单 11-1 所示。

代码清单 11-1　Hadoop JMX callback

```
hbasepractice({
  "beans" : [ {
    "name" : "java.lang:type=Memory",
    "modelerType" : "sun.management.MemoryImpl",
    "Verbose" : false,
    "HeapMemoryUsage" : {
      "committed" : 812646400,
      "init" : 2112558912,
      "max" : 12012486656,
      "used" : 552718016
    },
    "NonHeapMemoryUsage" : {
      "committed" : 59179008,
      "init" : 24576000,
      "max" : 136314880,
      "used" : 46579504
    },
    "ObjectPendingFinalizationCount" : 0,
    "ObjectName" : "java.lang:type=Memory"
  }, {
    "name" : "java.lang:type=MemoryPool,name=PS Eden Space",
    "modelerType" : "sun.management.MemoryPoolImpl",
    "CollectionUsage" : {
      "committed" : 173015040,
      "init" : 528482304,
      "max" : 4496818176,
      "used" : 0
    },
    "CollectionUsageThreshold" : 0,
    "CollectionUsageThresholdCount" : 0,
    "MemoryManagerNames" : [ "PS MarkSweep", "PS Scavenge" ],
    "PeakUsage" : {
      "committed" : 3999793152,
      "init" : 528482304,
      "max" : 4502585344,
```

```
      "used" : 3999793152
    },
    "Usage" : {
      "committed" : 173015040,
      "init" : 528482304,
      "max" : 4496818176,
      "used" : 150651592
    },
    "CollectionUsageThresholdExceeded" : false,
    "CollectionUsageThresholdSupported" : true,
    "UsageThresholdSupported" : false,
    "Name" : "PS Eden Space",
    "Type" : "HEAP",
    "Valid" : true,
    "ObjectName" : "java.lang:type=MemoryPool,name=PS Eden Space"
  },{
    ...
  } ]
});
```

（2）qry：Hadoop JMX 监控接口默认返回了所有的 JMX 监控信息。如果需要针对某个特定的 JMX 项做监控（如 JVM 内存使用情况），则 qry 请求就有了用武之地，例如接口 http://master1:50070/jmx?qry=java.lang:type=Memory 可以查询 NameNode JVM 内存使用情况。注意，qry 参数的值就是 JMX bean 的 name 值，请求返回数据如代码清单 11-2 所示。

代码清单 11-2　Hadoop JMX qry

```
{
  "beans" : [ {
    "name" : "java.lang:type=Memory",
    "modelerType" : "sun.management.MemoryImpl",
    "Verbose" : false,
    "HeapMemoryUsage" : {
      "committed" : 740294656,
      "init" : 2112558912,
      "max" : 12012486656,
      "used" : 336624200
    },
    "NonHeapMemoryUsage" : {
      "committed" : 57606144,
      "init" : 24576000,
      "max" : 136314880,
      "used" : 46655256
    },
```

```
      "ObjectPendingFinalizationCount" : 0,
      "ObjectName" : "java.lang:type=Memory"
  } ]
}
```

（3）get：如果需要更加细粒度地监控某个 JMX 项，则可以使用 get 参数，例如 http://master1:50070/jmx?get=java.lang:type=Memory::HeapMemoryUsage 可以获得 NameNode JVM 内存的堆内存使用情况，请求返回数据如代码清单 11-3 所示。

代码清单 11-3　Hadoop JMX get

```
{
  "beans" : [ {
    "name" : "java.lang:type=Memory",
    "modelerType" : "sun.management.MemoryImpl",
    "HeapMemoryUsage" : {
      "committed" : 738197504,
      "init" : 2112558912,
      "max" : 12012486656,
      "used" : 326854912
    }
  } ]
}
```

11.2　HBase 监控

国内诸如阿里、小米等大公司通常会采用 Alimonitor、open-falcon、ganglia 等软件监控 HBase 集群各节点机器的指标，通过自定义的服务定时拉取 Hadoop、HBase JMX 相关数据并持久化，最后通过一个自定义的控制台展示这些指标，并且集成企业内部的监控报警系统用来预警。本章介绍基础的 HBase Web 页面、JMX 与客户端监控。

11.2.1　Web 监控页面

HBase Web 监控页面与 Hadoop Web 监控页面相比可看到的监控信息大大增加，包括 HRegionServer 的启动情况、内存使用情况、每秒请求数（QPS）、每个分区服务器的分区负载、每个分区服务器正在执行的任务、HBase 表分区运行情况、每个分区的 StoreFile 文件个数与大小等。监控地址为 http://masterhost:16010。

该页面为集群概览监控页面，主要包括以下信息。

（1）节点列表与负载，如图 11-3 所示，监控信息包括：提供服务的分区服务器、已经死亡的分区服务器、每个分区服务器的内存使用情况、请求数与服务的分区数等。

11.2 HBase 监控

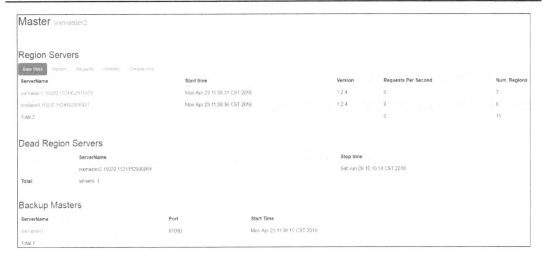

图 11-3 HBase 监控——集群节点

（2）集群表概览，如图 11-4 所示，包括用户表、系统表和快照列表。

图 11-4 HBase 监控——表概览

（3）运行任务，如图 11-5 所示，包括当前集群正在运行的任务与集群软件相关信息。

图 11-5 HBase 监控——运行任务与集群信息

（4）节点监控概览，如图 11-6 所示，包括当前节点负责的分区数、数据块本地化率、内存、请求、WAL 相关（数目与文件大小）、缓存使用情况以及最重要的当前运行任务监控。图 11-6 中任务列表为表 s_behavior 列族 pc 压缩（或者说合并）刚刚执行完成，如果任务正在执行就会显示一个任务的执行时长，例如，一个较大的分区执行一次主压缩需要半个小时，此时就需要注意该任务是否会对在线服务产生影响了。

图 11-6　HBase 监控——节点 Metrics

（5）分区监控，如图 11-7 所示，包括分区基本信息、请求 Metric、StoreFile Metric 等。需要注意的是 StoreFile Metric，如果某个 StoreFile 的大小达到了一定的条件（如超过 20 GB），可能就需要对这个 Store 对应的分区做拆分了。

图 11-7　HBase 监控——节点分区监控

11.2.2　JMX 监控

HBase JMX 接口使用方式与 Hadoop JMX 接口使用方式类似，同样返回所有的 JMX 信息，可以通过 qry、get 参数来限制需要查询的结果，HBase JMX 监控接口地址为：

```
HMaster: http://hmasterhost:16010/jmx
HRegionServer: http://hregionserverhost:16030/jmx
```

下面为几个查询示例。
- HMaster IPC 请求统计：http://master1:16010/jmx?qry=Hadoop:service=HBase,name=Master,sub=IPC。
- HMaster GC 信息：http://master1:16010/jmx?qry=java.lang:type=GarbageCollector,name=ConcurrentMarkSweep。
- 集群存活节点信息：http://master1:16010/jmx?get=Hadoop:service=HBase,name=Master,sub=Server::tag.liveRegionServers。

表 11-1 列出了一些 HBase JMX 支持的指标。

表 11-1 常用的集群间数据复制以及数据同步的方式

监控指标	作用域	描述
`java.lang:type=Memory`	本机	本机内存使用，包括堆与非堆内存
`Hadoop:service=HBase, name=Master, sub=IPC`	HMaster	IPC 请求统计相关，包括请求时间与返回值大小等
`Hadoop:service=HBase, name=RegionServer, sub=IPC`	HRegionServer	IPC 请求统计相关，包括请求时间与返回值大小等
`java.lang:type=GarbageCollector, name=ConcurrentMarkSweep`	本机	JVM GC 统计信息
`java.lang:type=MemoryPool, name=Par Eden Space`	本机	JVM Eden 区（堆）使用空间统计
`java.lang:type=MemoryPool, name=CMS Perm Gen`	本机	JVM Perm 区（非堆）使用空间统计
`java.lang:type=MemoryPool, name=Par Survivor Space`	本机	JVM Survivor 区（堆）使用空间统计
`java.lang:type=MemoryPool, name=CMS Old Gen`	本机	JVM Old Gen（堆）使用空间统计
`java.lang:type=GarbageCollector, name=ParNew`	本机	JVM ParNew（堆）使用空间统计
`Hadoop:service=HBase, name=RegionServer, sub=Replication`	HRegionServer	HBase 复制相关信息统计
`Hadoop:service=HBase, name=RegionServer, sub=Regions`	HRegionServer	复制的分区相关信息统计，包括 Store 数量、StoreFile 数量、读写请求数等
`Hadoop:service=HBase, name=RegionServer, sub=WAL`	HRegionServer	WAL 相关信息统计，包括刷新到磁盘时间统计等
`Hadoop:service=HBase, name=RegionServer, sub=Server`	HRegionServer	HRegionServer 服务信息统计，包括 Scan、Delete 等请求时长统计等

11.2.3　API 监控

HBase JMX 提供了最为丰富的监控信息，一般企业监控会通过定时使用 Http 请求 JMX 接口获得监控信息后持久化存储，并且结合企业内部的监控系统做预警，同时 HBase 也提供了客户端 API 用来监控一些 HBase 集群的基本信息（相对 JMX 监控信息，这些基本信息少得可怜），代码清单 11-4 演示了如何使用 HBase 客户端 API 来获取 HBase 集群在线的 HRegionServer 节点以及已经死亡的 HRegionServer 节点等信息。

代码清单 11-4　HBase API 监控

```
1.   package com.mt.hbase.chpt11.monitor;
2.
3.   import com.mt.hbase.connection.HBaseConnectionFactory;
4.   import org.apache.hadoop.hbase.ClusterStatus;
5.   import org.apache.hadoop.hbase.client.Admin;
6.   import org.apache.hadoop.hbase.client.HBaseAdmin;
7.
8.   import java.io.IOException;
9.
10.  public class HBaseMonitor {
11.
12.
13.      public static void main(String [] args) throws IOException {
14.
15.          Admin admin = HBaseConnectionFactory.getConnection().getAdmin();
16.
17.          HBaseAdmin hBaseAdmin = (HBaseAdmin) admin;
18.
19.          ClusterStatus clusterStatus = hBaseAdmin.getClusterStatus();
20.
21.          /**
22.           * 获得集群平均分区加载，即平均每台分区服务器负责几个分区
23.           */
24.          System.out.println(clusterStatus.getAverageLoad());
25.          /**
26.           * 获得集群在线的分区服务器
27.           */
28.          System.out.println(clusterStatus.getServers());
29.
30.          /**
31.           * 获得集群死亡的分区服务器
32.           */
33.          System.out.println(clusterStatus.getDeadServers());
34.
35.      }
```

```
36.  }
37.
```

  监控也是一个不断发展和积累的过程，不可能一开始就设计得很完善。监控为系统问题排查、故障预警提供了基础保障，不同的业务有着不同的监控需求，当业务比较简单时可以选用一些开源易用的监控软件，只有当业务已经复杂到需要定制才能解决问题时，才考虑开发自己独有的监控系统。

后记

本书从 2017 年 8 月开始动笔，写到这里已经接近尾声。刚开始我是以一种写技术博客和分享的心态去写的，后来在朋友的鼓励下才决定出版成书。在本书写作和出版过程中，人民邮电出版社杨海玲编辑给了我很多中肯的建议和意见，她的帮助和鼓励坚定了我出这本书的信念。

记得自己最开始将 HBase 应用到在线系统的时候，HBase 集群经常响应延迟、分区服务器宕机，最后不得不对业务做降级、限流以便减轻 HBase 集群压力而恢复服务。本书也将这些 HBase 在线调优的经验沉淀了下来。

写到最后虽然已经完成了自己刚开始定下的目标，但是仍然觉得还有很多知识点可以继续写下去，如在集群中规划不同资源需求型的业务、对业务透明二级索引的构建、权限控制、HBase 2.0 的一些新特性等，如果有机会希望能够在以后的更新版本中加入这些知识。

在写书的一年多时间内，我经历了从魅族较为稳定的生活到创业可能朝不保夕的风险，同时家庭的经济压力也较大，但是不管怎样，人生在于拼搏，因此我决定出来奋斗，希望有一天能够很笃定地对自己说，谢谢曾经努力的自己！

附录

常见问题

本附录将汇总一些 HBase 集群线上环境可能遇到的问题，分析问题产生的原因，并给出解决方案，开发人员或者运维人员在线上遇到类似问题时可以作为参考。

A.1 GC 时间超长导致分区服务器宕机

在我多年的一线经验当中，该问题是 HBase 线上出现最频繁的问题之一，HBase 分区服务器启动一段时间后自动下线，HBase 日志文件输出代码清单 A-1 所示的日志。

代码清单 A-1　HBase GC 超时导致分区服务器下线

```
2018-02-23 08:55:35,026 WARN  [JvmPauseMonitor] util.JvmPauseMonitor: Detected pause in JVM or host machine (eg GC): pause of approximately 25883ms
GC pool 'ParNew' had collection(s): count=1 time=26192ms
2018-02-23 08:55:35,067 INFO  [main-SendThread(master2:2181)] zookeeper.ClientCnxn: Client session timed out, have not heard from server in 27352ms for sessionid 0x25b220abf3d03d4, closing socket connection and attempting reconnect
2018-02-23 08:55:35,067 INFO  [pool-23-thread-1560-SendThread(nslave3:2181)] zookeeper.ClientCnxn: Client session timed out, have not heard from server in 35580ms for sessionid 0x35ed0bff76b020d, closing socket connection and attempting reconnect
2018-02-23 08:55:35,084 INFO  [ReplicationRpcServer.handler=1,queue=0,port=16020-SendThread(master1:2181)] zookeeper.ClientCnxn: Client session timed out, have not heard from server in 31963ms for sessionid 0x25b220abf3d03d5, closing socket connection and attempting reconnect
2018-02-23 08:55:35,084 INFO
……
2018-02-23 08:55:54,044 WARN  [regionserver/slave7/172.168.0.7:16020] util.Sleeper: We slept 19029ms instead of 3000ms, this is likely due to a long garbage collecting pause and it's usually bad, see http://hbase.apache.org/book.html#trouble.rs.runtime.zkexpired
```

2018-02-23 08:55:54,044 WARN [JvmPauseMonitor] util.JvmPauseMonitor: Detected pause in JVM or host machine (eg GC): pause of approximately 18018ms
GC pool 'ParNew' had collection(s): count=1 time=16956ms
2018-02-23 08:55:54,044 INFO [main-SendThread(master1:2181)] zookeeper.ClientCnxn: Opening socket connection to server master1/10.10.32.37:2181. Will not attempt to authenticate using SASL (unknown error)
2018-02-23 08:55:54,044 INFO [MemStoreFlusher.1] regionserver.HRegion: Flushing 1/1 column families, memstore=12.49 MB
2018-02-23 08:55:54,045 INFO [main-SendThread(master1:2181)] zookeeper.ClientCnxn: Socket connection established to master1/10.10.32.37:2181, initiating session
2018-02-23 08:55:54,045 INFO [slave7,16020,1506416633339_ChoreService_1] hbase.ScheduledChore: Chore: CompactionChecker missed its start time.
2018-02-23 08:55:54,045 INFO [slave7,16020,1506416633339_ChoreService_3] hbase.ScheduledChore: Chore: slave7,16020,1506416633339-MemstoreFlusherChore missed its start time
2018-02-23 08:55:54,045 INFO [slave7,16020,1506416633339_ChoreService_3] regionserver.HRegionServer: slave7,16020,1506416633339-MemstoreFlusherChore requesting flush of s_sms,076,1505873369585.1e98cb109d37bb727b99e17e6f4d714a. because cf h as an old edit so flush to free WALs after random delay 271431ms
2018-02-23 08:55:54,045 INFO [slave7,16020,1506416633339_ChoreService_3] regionserver.HRegionServer: slave7,16020,1506416633339-MemstoreFlusherChore requesting flush of s_call,026,1506069360055.fa8d5425881e37c47f7dd4febfbfcc0d. because cf h as an old edit so flush to free WALs after random delay 276901ms
2018-02-23 08:55:54,045 INFO [slave7,16020,1506416633339_ChoreService_3] regionserver.HRegionServer: slave7,16020,1506416633339-MemstoreFlusherChore requesting flush of s_sms,132,1505530492628.2d744d9ca8d5c7aa852fc02de793ea7b. because cf has an old edit so flush to free WALs after random delay 22589ms
2018-02-23 08:55:54,045 WARN [B.defaultRpcServer.handler=104,queue=4,port=16020] ipc.RpcServer: (responseTooSlow): {"processingtimems":19033,"call":"Scan(org.apache.hadoop.hbase.protobuf.generated.ClientProtos$ScanRequest)","client":"10.10.5.107:4463","starttimems":1519347335012,"queuetimems":5,"class":"HRegionServer","responsesize":11,"method":"Scan"}
2018-02-23 08:55:54,045 WARN [B.defaultRpcServer.handler=78,queue=18,port=16020] ipc.RpcServer: (responseTooSlow): {"processingtimems":19000,"call":"Scan(org.apache.hadoop.hbase.protobuf.generated.ClientProtos$ScanRequest)","client":"10.10.5.101:25773","starttimems":1519347335045,"queuetimems":22,"class":"HRegionServer","responsesize":11,"method":"Scan"}
2018-02-23 08:55:54,045 INFO [slave7,16020,1506416633339_ChoreService_3] regionserver.HRegionServer: slave7,16020,1506416633339-MemstoreFlusherChore requesting flush of s_call,504,1506084570443.676bb2c2986e6f0997ec967b2132a342. because cf h as an old edit so flush to free WALs after random delay 30523ms
2018-02-23 08:55:54,045 WARN [B.defaultRpcServer.handler=35,queue=15,port=16020] ipc.RpcServer: (responseTooSlow): {"processingtimems":18935,"call":"Scan(org.apache.hadoop.hbase.protobuf.generated.ClientProtos$ScanRequest)","client":"10.10.5.106:31042","starttimems":1519347335110,"queuetimems":86,"class":"HRegionServer","responsesize":11,"method":"Scan"}
……
2018-02-23 08:55:54,049 FATAL [main-EventThread] regionserver.HRegionServer: ABORTING region server slave7,16020,1506416633339: regionserver:16020-0x25b220abf3d

03d4, quorum=master1:2181,master2:2181,slave1:2181, baseZNode=/hbase regionserver:16020-0x25b220abf3d03d4 received expired from ZooKeeper, aborting
org.apache.zookeeper.KeeperException$SessionExpiredException: KeeperErrorCode = Session expired
    at org.apache.hadoop.hbase.zookeeper.ZooKeeperWatcher.connectionEvent(ZooKeeperWatcher.java:702)
    at org.apache.hadoop.hbase.zookeeper.ZooKeeperWatcher.process(ZooKeeperWatcher.java:613)
    at org.apache.zookeeper.ClientCnxn$EventThread.processEvent(ClientCnxn.java:522)
    at org.apache.zookeeper.ClientCnxn$EventThread.run(ClientCnxn.java:498)
2018-02-23 08:55:54,222 WARN   [B.defaultRpcServer.handler=133,queue=13,port=16020] ipc.RpcServer: (responseTooSlow): {"processingtimems":19215,"call":"Scan(org.apache.hadoop.hbase.protobuf.generated.ClientProtos$ScanRequest)","client":"10.10.5.98:62342","starttimems":1519347335007,"queuetimems":2,"class":"HregionServer","responsesize":11,"method":"Scan"}
2018-02-23 08:55:54,221 WARN   [DataStreamer for file /user/hbase/WALs/slave7,16020,1506416633339/slave7%2C16020%2C1506416633339.default.1519347230724 block BP-1729951490-10.10.32.37-1490924840039:blk_1092226654_18488799] hdfs.DFSClient: DataStreamer Exception
java.io.IOException: Broken pipe
    at sun.nio.ch.FileDispatcherImpl.write0(Native Method)
    at sun.nio.ch.SocketDispatcher.write(SocketDispatcher.java:47)
    at sun.nio.ch.IOUtil.writeFromNativeBuffer(IOUtil.java:93)
    at sun.nio.ch.IOUtil.write(IOUtil.java:65)
    at sun.nio.ch.SocketChannelImpl.write(SocketChannelImpl.java:487)
    at org.apache.hadoop.net.SocketOutputStream$Writer.performIO(SocketOutputStream.java:63)
    at org.apache.hadoop.net.SocketIOWithTimeout.doIO(SocketIOWithTimeout.java:142)
    at org.apache.hadoop.net.SocketOutputStream.write(SocketOutputStream.java:159)
    at org.apache.hadoop.net.SocketOutputStream.write(SocketOutputStream.java:117)
    at java.io.BufferedOutputStream.flushBuffer(BufferedOutputStream.java:82)
    at java.io.BufferedOutputStream.flush(BufferedOutputStream.java:140)
    at java.io.DataOutputStream.flush(DataOutputStream.java:123)
    at org.apache.hadoop.hdfs.DFSOutputStream$DataStreamer.run(DFSOutputStream.java:581)
2018-02-23 08:55:54,219 WARN   [B.defaultRpcServer.handler=194,queue=14,port=16020] ipc.RpcServer: (responseTooSlow): {"processingtimems":19130,"call":"Scan(org.apache.hadoop.hbase.protobuf.generated.ClientProtos$ScanRequest)","client":"10.10.5.107:4463","starttimems":1519347335089,"queuetimems":82,"class":"HregionServer","responsesize":11,"method":"Scan"}
2018-02-23 08:55:54,190 FATAL [regionserver/slave7/172.168.0.7:16020.replicationSource,3.replicationSource.slave7%2C16020%2C1506416633339.default,3] regionserver.HRegionServer: ABORTING region server slave7,16020,1506416633339: Failed to write replication wal position (filename=slave7%2C16020%2C1506416633339.default.1519347230724, position=40174334)
org.apache.zookeeper.KeeperException$SessionExpiredException: KeeperErrorCode = Session expired for /hbase/replication/rs/slave7,16020,1506416633339/3/slave7%2C16020%2C1506416633339.default.1519347230724
    at org.apache.zookeeper.KeeperException.create(KeeperException.java:127)
    at org.apache.zookeeper.KeeperException.create(KeeperException.java:51)

```
        at org.apache.zookeeper.ZooKeeper.setData(ZooKeeper.java:1270)
        at org.apache.hadoop.hbase.zookeeper.RecoverableZooKeeper.setData(Recoverabl
eZooKeeper.java:422)
        at org.apache.hadoop.hbase.zookeeper.ZKUtil.setData(ZKUtil.java:818)
        at org.apache.hadoop.hbase.zookeeper.ZKUtil.setData(ZKUtil.java:868)
        at org.apache.hadoop.hbase.zookeeper.ZKUtil.setData(ZKUtil.java:862)
        at org.apache.hadoop.hbase.replication.ReplicationQueuesZKImpl.setLogPositio
n(ReplicationQueuesZKImpl.java:130)
        at org.apache.hadoop.hbase.replication.regionserver.ReplicationSourceManager.
logPositionAndCleanOldLogs(ReplicationSourceManager.java:184)
        at org.apache.hadoop.hbase.replication.regionserver.ReplicationSource$Replic
ationSourceWorkerThread.shipEdits(ReplicationSource.java:940)
        at org.apache.hadoop.hbase.replication.regionserver.ReplicationSource$Replic
ationSourceWorkerThread.run(ReplicationSource.java:620)
2018-02-23 08:55:54,190 WARN   [B.defaultRpcServer.handler=168,queue=8,port=16020]
ipc.RpcServer: (responseTooSlow): {"processingtimems":18671,"call":"Scan(org.apa
che.hadoop.hbase.protobuf.generated.ClientProtos$ScanRequest)","client":"10.10.5.
105:53764","starttimems":1519347335518,"queuetimems":459,"class":"HRegionServer",
"responsesize":15,"method":"Scan"}
2018-02-23 08:55:54,115 WARN   [ResponseProcessor for block BP-1729951490-10.10.3
2.37-1490924840039:blk_1092226654_18488799] hdfs.DFSClient: DFSOutputStream Resp
onseProcessor exception  for block BP-1729951490-10.10.32.37-1490924840039:blk_
1092226654_18488799
java.io.EOFException: Premature EOF: no length prefix available
        at org.apache.hadoop.hdfs.protocolPB.PBHelper.vintPrefixed(PBHelper.java:2000)
        at org.apache.hadoop.hdfs.protocol.datatransfer.PipelineAck.readFields(Pipel
ineAck.java:176)
        at org.apache.hadoop.hdfs.DFSOutputStream$DataStreamer$ResponseProcessor.run
(DFSOutputStream.java:798)
2018-02-23 08:55:54,109 WARN   [B.defaultRpcServer.handler=139,queue=19,port=16020]
ipc.RpcServer: (responseTooSlow): {"processingtimems":18752,"call":"Scan(org.apache.
hadoop.hbase.protobuf.generated.ClientProtos$ScanRequest)","client":"10.10.5.98:
62342","starttimems":1519347335357,"queuetimems":0,"class":"HRegionServer","resp
onsesize":15,"method":"Scan"}
2018-02-23 08:55:54,462 WARN   [DataStreamer for file /user/hbase/WALs/slave7,160
20,1506416633339/slave7%2C16020%2C1506416633339.default.1519347230724 block BP-1
729951490-10.10.32.37-1490924840039:blk_1092226654_18488799] hdfs.DFSClient: Err
or Recovery for block BP-1729951490-10.10.32.37-1490924840039:blk_1092226654_184
88799 in pipeline 172.168.0.7:50010, 10.10.32.43:50010, 10.10.32.26:50010: bad d
atanode 172.168.0.7:50010
2018-02-23 08:55:54,095 WARN   [B.defaultRpcServer.handler=121,queue=1,port=16020]
ipc.RpcServer: (responseTooSlow): {"processingtimems":18543,"call":"Scan(org.apa
che.hadoop.hbase.protobuf.generated.ClientProtos$ScanRequest)","client":"10.10.5.
98:62342","starttimems":1519347335551,"queuetimems":522,"class":"HRegionServer",
"responsesize":15,"method":"Scan"}

2018-02-23 08:55:54,061 INFO   [pool-23-thread-1560-SendThread(nslave2:2181)] zoo
keeper.ClientCnxn: Opening socket connection to server nslave2/10.10.234.190:218
1. Will not attempt to authenticate using SASL (unknown error)
```

```
2018-02-23 08:55:54,060 INFO  [ReplicationRpcServer.handler=2,queue=0,port=16020]
client.AsyncProcess: #3, waiting for 100000  actions to finish on table: s_behavior
......
2018-02-23 08:55:54,524 INFO  [pool-23-thread-1560-SendThread(nslave2:2181)] zoo
keeper.ClientCnxn: Unable to reconnect to ZooKeeper service, session 0x35ed0bff7
6b020d has expired, closing socket connection
2018-02-23 08:55:54,524 ERROR [pool-23-thread-1560-EventThread] replication.Hbas
eReplicationEndpoint: The HBaseReplicationEndpoint corresponding to peer 3 was a
borted for the following reason(s):connection to cluster: 3-0x35ed0bff76b020d, q
uorum=nslave1:2181,nslave2:2181,nslave3:2181, baseZNode=/hbase connection to clu
ster: 3-0x35ed0bff76b020d received expired from ZooKeeper, aborting
org.apache.zookeeper.KeeperException$SessionExpiredException: KeeperErrorCode =
Session expired
    at org.apache.hadoop.hbase.zookeeper.ZooKeeperWatcher.connectionEvent(ZooKee
perWatcher.java:702)
    at org.apache.hadoop.hbase.zookeeper.ZooKeeperWatcher.process(ZooKeeperWatch
er.java:613)
    at org.apache.zookeeper.ClientCnxn$EventThread.processEvent(ClientCnxn.java:522)
    at org.apache.zookeeper.ClientCnxn$EventThread.run(ClientCnxn.java:498)
2018-02-23 08:55:54,524 INFO  [pool-23-thread-1560-EventThread] zookeeper.Client
Cnxn: EventThread shut down
2018-02-23 08:55:54,533 WARN  [B.defaultRpcServer.handler=160,queue=0,port=16020]
ipc.RpcServer: (responseTooSlow): {"processingtimems":19525,"call":"Scan(org.apa
che.hadoop.hbase.protobuf.generated.ClientProtos$ScanRequest)","client":"10.10.5.
101:25773","starttimems":1519347335007,"queuetimems":1,"class":"HRegionServer","
responsesize":11,"method":"Scan"}
......
2018-02-23 08:55:55,205 WARN  [B.defaultRpcServer.handler=157,queue=17,port=16020]
regionserver.MultiVersionConcurrencyControl: STUCK: MultiVersionConcurrencyContr
ol{readPoint=38025361, writePoint=38025365}
2018-02-23 08:55:55,205 WARN  [B.defaultRpcServer.handler=27,queue=7,port=16020]
regionserver.MultiVersionConcurrencyControl: STUCK: MultiVersionConcurrencyContr
ol{readPoint=38025361, writePoint=38025365}
2018-02-23 08:55:55,271 WARN  [hconnection-0x66ef393c-shared--pool3-t3738200]
client.AsyncProcess: #3, table=s_call, attempt=4/4 failed=20ops, last exception:
org.apache.hadoop.hbase.regionserver.RegionServerAbortedException: org.apache.ha
doop.hbase.regionserver.RegionServerAbortedException: Server slave7,16020,150641
6633339 aborting
    at org.apache.hadoop.hbase.regionserver.RSRpcServices.checkOpen(RSRpcServices.
java:1093)
    at org.apache.hadoop.hbase.regionserver.RSRpcServices.multi(RSRpcServices.
java:2078)
    at org.apache.hadoop.hbase.protobuf.generated.ClientProtos$ClientService$2.
callBlockingMethod(ClientProtos.java:33656)
    at org.apache.hadoop.hbase.ipc.RpcServer.call(RpcServer.java:2180)
    at org.apache.hadoop.hbase.ipc.CallRunner.run(CallRunner.java:112)
    at org.apache.hadoop.hbase.ipc.RpcExecutor.consumerLoop(RpcExecutor.java:133)
    at org.apache.hadoop.hbase.ipc.RpcExecutor$1.run(RpcExecutor.java:108)
    at java.lang.Thread.run(Thread.java:745)
```

```
on slave7,16020,1506416633339, tracking started Fri Feb 23 08:55:54 CST 2018;
not retrying 20 - final failure
2018-02-23 08:55:55,414 FATAL [regionserver/slave7/172.168.0.7:16020] regionserver.
HRegionServer: ABORTING region server slave7,16020,1506416633339: org.apache.had
oop.hbase.YouAreDeadException: Server REPORT rejected; currently processing slav
e7,16020,1506416633339 as dead server
        at org.apache.hadoop.hbase.master.ServerManager.checkIsDead(ServerManager.
java:434)
        at org.apache.hadoop.hbase.master.ServerManager.regionServerReport(ServerMan
ager.java:339)
        at org.apache.hadoop.hbase.master.MasterRpcServices.regionServerReport(Maste
rRpcServices.java:339)
        at org.apache.hadoop.hbase.protobuf.generated.RegionServerStatusProtos$Regio
nServerStatusService$2.callBlockingMethod(RegionServerStatusProtos.java:8617)
        at org.apache.hadoop.hbase.ipc.RpcServer.call(RpcServer.java:2180)
        at org.apache.hadoop.hbase.ipc.CallRunner.run(CallRunner.java:112)
        at org.apache.hadoop.hbase.ipc.RpcExecutor.consumerLoop(RpcExecutor.java:133)
        at org.apache.hadoop.hbase.ipc.RpcExecutor$1.run(RpcExecutor.java:108)
        at java.lang.Thread.run(Thread.java:745)
```

HBase 分区服务器 GC 日志片段如代码清单 A-2 所示。

代码清单 A-2　HRegionServer GC 日志

```
2018-02-23T08:55:08.811+0800: 12930678.090: [GC2018-02-23T08:55:08.812+0800: 129
30678.091: [ParNew
Desired survivor size 429490176 bytes, new threshold 1 (max 8)
- age   1:  447761072 bytes,  447761072 total
- age   2:  180667128 bytes,  628428200 total
- age   3:    8095608 bytes,  636523808 total
- age   4:    9067848 bytes,  645591656 total
- age   5:   17137912 bytes,  662729568 total
- age   6:    8682736 bytes,  671412304 total
- age   7:   19797800 bytes,  691210104 total
- age   8:   30364256 bytes,  721574360 total
: 2749409K->726602K(3355456K), 26.1922590 secs] 19074448K->17057853K(24326976K),
 26.1929470 secs] [Times: user=121.09 sys=1.24, real=26.19 secs]
2018-02-23T08:55:37.052+0800: 12930706.331: [GC2018-02-23T08:55:37.052+0800: 129
30706.331: [ParNew
Desired survivor size 429490176 bytes, new threshold 8 (max 8)
- age   1:  339365184 bytes,  339365184 total
: 3243103K->563128K(3355456K), 16.9557420 secs] 19574354K->17448922K(24326976K),
 16.9563520 secs] [Times: user=355.42 sys=2.10, real=16.95 secs]
2018-02-23T08:55:54.020+0800: 12930723.299: [GC [1 CMS-initial-mark: 16885793K(2
0971520K)] 17449204K(24326976K), 0.0231490 secs] [Times: user=0.02 sys=0.00, rea
l=0.03 secs]
2018-02-23T08:55:54.044+0800: 12930723.323: [CMS-concurrent-mark-start]
2018-02-23T08:55:55.467+0800: 12930724.746: [CMS-concurrent-mark: 1.245/1.423 se
cs] [Times: user=11.02 sys=0.43, real=1.42 secs]
2018-02-23T08:55:55.467+0800: 12930724.746: [CMS-concurrent-preclean-start]
```

```
2018-02-23T08:55:55.600+0800: 12930724.879: [CMS-concurrent-preclean: 0.130/0.133
secs] [Times: user=0.27 sys=0.02, real=0.14 secs]
2018-02-23T08:55:55.601+0800: 12930724.880: [CMS-concurrent-abortable-preclean
-start]
 CMS: abort preclean due to time 2018-02-23T08:56:00.668+0800: 12930729.947:
[CMS-concurrent-abortable-preclean: 4.119/5.067 secs] [Times: user=7.56 sys=
0.38, real=5.06 secs]
2018-02-23T08:56:00.681+0800: 12930729.960: [GC[YG occupancy: 2942653 K (3355456 K)]
2018-02-23T08:56:00.681+0800: 12930729.960: [Rescan (parallel) , 0.3055900 secs]
2018-02-23T08:56:00.987+0800: 12930730.266: [weak refs processing, 0.1313730 secs]
2018-02-23T08:56:01.118+0800: 12930730.397: [scrub string table, 0.0012490 secs]
[1 CMS-remark: 16885793K(20971520K)] 19828447K(24326976K), 0.4505510 secs] [Times:
user=7.08 sys=0.03, real=0.45 secs]
2018-02-23T08:56:01.132+0800: 12930730.411: [CMS-concurrent-sweep-start]
2018-02-23T08:56:05.073+0800: 12930734.352: [CMS-concurrent-sweep: 3.928/3.942
secs] [Times: user=4.43 sys=0.05, real=3.94 secs]
2018-02-23T08:56:05.073+0800: 12930734.352: [CMS-concurrent-reset-start]
2018-02-23T08:56:05.134+0800: 12930734.413: [CMS-concurrent-reset: 0.060/0.060
secs] [Times: user=0.06 sys=0.00, real=0.06 secs]
Heap
 par new generation    total 3355456K, used 2994605K [0x00000001d8000000, 0x00000
002d8000000, 0x00000002d8000000)
  eden space 2516608K,  96% used [0x00000001d8000000, 0x000000026c67d378, 0x0000
0002719a0000)
  from space 838848K,  67% used [0x00000002719a0000, 0x0000000293f8e108, 0x00000
002a4cd0000)
  to   space 838848K,   0% used [0x00000002a4cd0000, 0x00000002a4cd0000, 0x00000
002d8000000)
 concurrent mark-sweep generation total 20971520K, used 6603587K [0x00000002d800
0000, 0x00000007d8000000, 0x00000007d8000000)
 concurrent-mark-sweep perm gen total 131072K, used 49016K [0x00000007d8000000,
0x00000007e0000000, 0x0000000800000000)
```

注意代码清单 A-1 中高亮的两行日志，JVM 两次 GC 时间 Stop-The-World 停顿了超过 40 秒，之后分区服务器被 ZooKeeper 认为已经超时死亡，将节点从集群移除，并对该分区服务器负责的分区进行迁移，最后该分区服务器被停止。

这是一台 32 核的服务器，其中 HRegionServer 的 JVM 配置参数如代码清单 A-3 所示，可以看到堆内存为 24 GB，其中新生代大小为 4 GB，GC 日志中 Stop-The-World 耗时正是来自新生代的 ParNew GC，因此问题很可能是由于新生代内存配置过大导致 ParNew GC 所需时间较长，同样老生代内存为 20 GB，Full GC 也可能由于停顿时间过长而导致分区服务器与 ZooKeeper 会话过期而下线。

该问题可以从两个方向去解决。

- 修改 ZooKeeper 超时时间：参数 `zookeeper.session.timeout` 表示分区服务器与 ZooKeeper 的连接超时时间。在超时时间内分区服务器没有发生心跳给 ZooKeeper，分区服务器就会被 ZooKeeper 从集群中移除。若将该时间适当调长就可以缓解由于 GC 时间过长而导致会话过期的问题,但是时间过长也会导致分区服务器宕机后集群的容灾时间过长,即需等待会话过期后宕机的分区服务器所负责的分区才会被重新均衡指派给其他分区服务器负责。

- GC 调优:适当减少新生代内存大小,例如从 4 GB 减少到 2 GB。当然整个 HBase 的堆内存也可以调优,将一个大内存的 HBase 分区服务器节点拆分为两个或者多个小内存的节点(每个节点堆内存不超过 16 GB),具体可参考第 9 章的 JVM 内存调优章节。该方案基本没有负面影响,而且是当 HBase 集群负载达到一定程度后必须要做的调优。

代码清单 A-3　HRegionServer JVM 配置参数

```
-XX:+UseParNewGC -XX:+UseConcMarkSweepGC -XX:CMSInitiatingOccupancyFraction=70
-Xss256k -Xmx24g -Xms24g -Xmn4g -XX:MaxDirectMemorySize=26g -XX:SurvivorRatio=3
-XX:+UseParNewGC -XX:+UseConcMarkSweepGC -XX:MaxTenuringThreshold=8
-XX:CMSInitiatingOccupancyFraction=80 -XX:+UseCMSCompactAtFullCollection
-XX:+UseCMSInitiatingOccupancyOnly
```

A.2　Scanner 租期过期

该问题表现为 HBase 整体请求响应延迟,日志文件输出大量日志,如代码清单 A-4 所示。

代码清单 A-4　HBase Scanner 租期过期

```
2018-01-02 10:11:50,156 INFO  [regionserver/slave6/10.10.234.154:16020.leaseChecker]
regionserver.RSRpcServices: Scanner 198642638 lease expired on region s_behavior,
809,1506133362489.fb38706619acfb8b62018c6034ff59a8.
2018-01-02 10:11:50,156 INFO  [regionserver/slave6/10.10.234.154:16020.leaseChecker]
regionserver.RSRpcServices: Scanner 198642641 lease expired on region s_behavior,
804,1506077977761.9c2ef2683093d38542ee902aba88bc50.
2018-01-02 10:11:50,156 INFO  [regionserver/slave6/10.10.234.154:16020.leaseChecker]
regionserver.RSRpcServices: Scanner 198642639 lease expired on region s_behavior,
876,1506219220357.0f9d42e2aa6f17944975c5abe76e271a.
2018-01-02 10:11:50,156 INFO  [regionserver/slave6/10.10.234.154:16020.leaseChecker]
regionserver.RSRpcServices: Scanner 198642642 lease expired on region s_behavior,
897,1506353029283.654a40c866f2a563d599bb621aa91872.
2018-01-02 10:11:50,157 INFO  [regionserver/slave6/10.10.234.154:16020.leaseChecker]
regionserver.RSRpcServices: Scanner 198642637 lease expired on region s_behavior,
844,1506267389280.3d598fa8faff4da8f6777f7e501dfa8e.
2018-01-02 10:11:50,157 INFO  [regionserver/slave6/10.10.234.154:16020.leaseChecker]
regionserver.RSRpcServices: Scanner 198642643 lease expired on region s_behavior,
894,1506354122194.78886f5d130690191397fd5759e97b4f.
2018-01-02 10:11:50,259 INFO  [regionserver/slave6/10.10.234.154:16020.leaseChecker]
regionserver.RSRpcServices: Scanner 198642644 lease expired on region s_behavior,
822,1506172449907.68be942dc1fd161646867ee7b84c9ddf.
```

HBase 为每个分区服务器维护了一个 Leases 列表,这个 Leases 列表记载了当前未完成的 Scanner 的租期，并且定时（默认 10 s）扫描这些未完成的 Scanner 租期是否到期。如果一个 Scanner 在配置的时间（配置项 `hbase.client.scanner.timeout.period`）内仍然未执行完成，则此时 Leases 会将 Scanner 做过期处理，同时分区服务器会将 Scanner 关闭，此时即会打印如代码清单 A-4 所示日志。这种情况下一般都是 HBase 分区服务器已经出现了性能问题，有可能是由于 CPU、磁盘读写等压力过大或者请求过多导致响应延迟，如果确定是此类问题就需要考虑扩容 HBase 集群了。

A.3 分区迁移异常

HBase 客户端大量报错，主要集中在两个异常，即 `NotServingRegionException` 和 `RegionMovedException`，客户端日志文件输出日志如代码清单 A-5 所示。

代码清单 A-5　HBase 分区迁移异常

```
Fri Jun 15 09:20:44 CST 2018, RpcRetryingCaller{globalStartTime=1529025642783,
pause=100, retries=5}, org.apache.hadoop.hbase.exceptions.RegionMovedException:
Region moved to: hostname=master2 port=16020 startCode=1529025427334. As of
locationSeqNum=129993820.
        at org.apache.hadoop.hbase.client.RpcRetryingCaller.callWithRetries
(RpcRetryingCaller.java:157)
        at org.apache.hadoop.hbase.client.HTable.increment(HTable.java:1132)
        at com.meizu.sync.hbase.HBaseClient.incr(HBaseClient.java:599)
        at com.meizu.sync.hbase.HBClientWrapper.incr(HBClientWrapper.java:179)
        at com.meizu.sync.capacity.handler.VolumeServiceImpl.updateVolumeToHBase
(VolumeServiceImpl.java:170)
        at com.meizu.sync.capacity.handler.VolumeServiceImpl.updateVolume(Volume
ServiceImpl.java:125)
        at com.meizu.sync.service.impl.SyncVolumeServiceImpl.updateVolume(Sync
VolumeServiceImpl.java:127)
        at com.meizu.sync.service.impl.AsyncVolumeCalculator$1.run(AsyncVolume
Calculator.java:60)
        at java.lang.Thread.run(Thread.java:722)
Caused by: org.apache.hadoop.hbase.exceptions.RegionMovedException: Region moved to:
 hostname=master2 port=16020 startCode=1529025427334. As of locationSeqNum=129993820.
        at sun.reflect.GeneratedConstructorAccessor342.newInstance(Unknown Source)
        at sun.reflect.DelegatingConstructorAccessorImpl.newInstance(Delegating
ConstructorAccessorImpl.java:45)
        at java.lang.reflect.Constructor.newInstance(Constructor.java:525)
        at org.apache.hadoop.ipc.RemoteException.instantiateException(RemoteExcep
tion.java:106)
        at org.apache.hadoop.ipc.RemoteException.unwrapRemoteException(RemoteExce
ption.java:95)
        at org.apache.hadoop.hbase.protobuf.ProtobufUtil.getRemoteException(Proto
bufUtil.java:329)
```

```
        at org.apache.hadoop.hbase.client.HTable$7.call(HTable.java:1128)
        at org.apache.hadoop.hbase.client.HTable$7.call(HTable.java:1116)
        at org.apache.hadoop.hbase.client.RpcRetryingCaller.callWithRetries(Rpc
RetryingCaller.java:136)
        ... 8 more
Caused by: org.apache.hadoop.hbase.ipc.RemoteWithExtrasException(org.apache.hado
op.hbase.exceptions.RegionMovedException): org.apache.hadoop.hbase.exceptions.
RegionMovedException: Region moved to: hostname=master2 port=16020 startCode=
1529025427334. As of locationSeqNum=129993820.
        at org.apache.hadoop.hbase.regionserver.HRegionServer.getRegionByEncoded
Name(HRegionServer.java:2913)
        at org.apache.hadoop.hbase.regionserver.RSRpcServices.getRegion(RSRpcSer
vices.java:1059)
        at org.apache.hadoop.hbase.regionserver.RSRpcServices.mutate(RSRpcServices.
java:2213)
        at org.apache.hadoop.hbase.protobuf.generated.ClientProtos$ClientService$2.
callBlockingMethod(ClientProtos.java:33646)
        at org.apache.hadoop.hbase.ipc.RpcServer.call(RpcServer.java:2180)
        at org.apache.hadoop.hbase.ipc.CallRunner.run(CallRunner.java:112)
        at org.apache.hadoop.hbase.ipc.RpcExecutor.consumerLoop(RpcExecutor.java:133)
        at org.apache.hadoop.hbase.ipc.RpcExecutor$1.run(RpcExecutor.java:108)
        at java.lang.Thread.run(Thread.java:745)

        at org.apache.hadoop.hbase.ipc.RpcClientImpl.call(RpcClientImpl.java:1267)
        at org.apache.hadoop.hbase.ipc.AbstractRpcClient.callBlockingMethod(Abstr
actRpcClient.java:227)
        at org.apache.hadoop.hbase.ipc.AbstractRpcClient$BlockingRpcChannelImplem
entation.callBlockingMethod(AbstractRpcClient.java:336)
        at org.apache.hadoop.hbase.protobuf.generated.ClientProtos$ClientService$
BlockingStub.mutate(ClientProtos.java:34082)
        at org.apache.hadoop.hbase.client.HTable$7.call(HTable.java:1125)
        ... 10 more

Fri Jun 15 09:21:04 CST 2018, RpcRetryingCaller{globalStartTime=1529025663488,
pause=100, retries=5}, **org.apache.hadoop.hbase.NotServingRegionException: org.
apache.hadoop.hbase.NotServingRegionException: Region s_behavior,40,1510555604783.
805167e17dad375b63d8e1a2ef77868c. is not online on slave4,16020,1506407427543**
        at org.apache.hadoop.hbase.regionserver.HRegionServer.getRegionByEncoded
Name(HRegionServer.java:2922)
        at org.apache.hadoop.hbase.regionserver.RSRpcServices.getRegion(RSRpcSer
vices.java:1059)
        at org.apache.hadoop.hbase.regionserver.RSRpcServices.mutate(RSRpcServices.
java:2213)
        at org.apache.hadoop.hbase.protobuf.generated.ClientProtos$ClientService$2.
callBlockingMethod(ClientProtos.java:33646)
        at org.apache.hadoop.hbase.ipc.RpcServer.call(RpcServer.java:2180)
        at org.apache.hadoop.hbase.ipc.CallRunner.run(CallRunner.java:112)
        at org.apache.hadoop.hbase.ipc.RpcExecutor.consumerLoop(RpcExecutor.java:133)
        at org.apache.hadoop.hbase.ipc.RpcExecutor$1.run(RpcExecutor.java:108)
```

```
        at java.lang.Thread.run(Thread.java:745)


        at org.apache.hadoop.hbase.client.RpcRetryingCaller.callWithRetries(RpcRe
tryingCaller.java:157)
        at org.apache.hadoop.hbase.client.HTable.increment(HTable.java:1132)
        at com.meizu.sync.hbase.HBaseClient.incr(HBaseClient.java:599)
        at com.meizu.sync.hbase.HBClientWrapper.incr(HBClientWrapper.java:179)
        at com.meizu.sync.capacity.handler.VolumeServiceImpl.updateVolumeToHBase
(VolumeServiceImpl.java:170)
        at com.meizu.sync.capacity.handler.VolumeServiceImpl.updateVolume(Volume
ServiceImpl.java:125)
        at com.meizu.sync.service.impl.SyncVolumeServiceImpl.updateVolume(SyncVol
umeServiceImpl.java:127)
        at com.meizu.sync.service.impl.AsyncVolumeCalculator$1.run(AsyncVolumeCal
culator.java:60)
        at java.lang.Thread.run(Thread.java:722)
Caused by: org.apache.hadoop.hbase.NotServingRegionException: org.apache.hadoop.
hbase.NotServingRegionException: Region s_behavior,40,1510555604783.805167e17dad
375b63d8e1a2ef77868c. is not online on slave4,16020,1506407427543
        at org.apache.hadoop.hbase.regionserver.HRegionServer.getRegionByEncoded
Name(HRegionServer.java:2922)
        at org.apache.hadoop.hbase.regionserver.RSRpcServices.getRegion(RSRpc
Services.java:1059)
        at org.apache.hadoop.hbase.regionserver.RSRpcServices.mutate(RSRpcServices.
java:2213)
        at org.apache.hadoop.hbase.protobuf.generated.ClientProtos$ClientService$2.
callBlockingMethod(ClientProtos.java:33646)
        at org.apache.hadoop.hbase.ipc.RpcServer.call(RpcServer.java:2180)
        at org.apache.hadoop.hbase.ipc.CallRunner.run(CallRunner.java:112)
        at org.apache.hadoop.hbase.ipc.RpcExecutor.consumerLoop(RpcExecutor.java:133)
        at org.apache.hadoop.hbase.ipc.RpcExecutor$1.run(RpcExecutor.java:108)
        at java.lang.Thread.run(Thread.java:745)
        at sun.reflect.GeneratedConstructorAccessor348.newInstance(Unknown Source)
        at sun.reflect.DelegatingConstructorAccessorImpl.newInstance(Delegating
ConstructorAccessorImpl.java:45)
        at java.lang.reflect.Constructor.newInstance(Constructor.java:525)
        at org.apache.hadoop.ipc.RemoteException.instantiateException(RemoteException.
java:106)
        at org.apache.hadoop.ipc.RemoteException.unwrapRemoteException(RemoteException.
java:95)
        at org.apache.hadoop.hbase.protobuf.ProtobufUtil.getRemoteException(Proto
bufUtil.java:329)
        at org.apache.hadoop.hbase.client.HTable$7.call(HTable.java:1128)
        at org.apache.hadoop.hbase.client.HTable$7.call(HTable.java:1116)
        at org.apache.hadoop.hbase.client.RpcRetryingCaller.callWithRetries(Rpc
RetryingCaller.java:136)
        ... 8 more
```

这两个异常都是由分区迁移引起的，RegionMovedException 表示分区正在迁移，

而 `NotServingRegionException` 则表示由于分区已经迁移，而客户端元数据缓存未更新，因此分区已经不是由当前分区服务器负责管理了。不过客户端会重试去获取最新的元数据，因此该异常并不会影响业务，而分区迁移通常发生在 HBase 分区服务器宕机或者新节点上线时，如果开启了自动负载均衡，则可能偶尔会有分区的迁移发生，总之该异常不会影响业务，但是需要检查是否有 HBase 分区服务器节点宕机。

A.4　Windows 运行 HBase 程序缺少 winutil.exe

很多公司开发人员使用的计算机操作系统都是 Windows 操作系统。当在 Windows 操作系统上执行 HBase 应用程序的时候，可能会出现如代码清单 A-6 所示的异常。

代码清单 A-6　Windows 缺少 winutil.exe

```
2018-01-01 16:03:13,212 ERROR [org.apache.hadoop.util.Shell] - Failed to locate the winutils binary in the hadoop binary path
java.io.IOException: Could not locate executable null\bin\winutils.exe in the Hadoop binaries.
    at org.apache.hadoop.util.Shell.getQualifiedBinPath(Shell.java:278)
    at org.apache.hadoop.util.Shell.getWinUtilsPath(Shell.java:300)
    at org.apache.hadoop.util.Shell.<clinit>(Shell.java:293)
```

因为 Hadoop 在 Windows 操作系统上运行的时候需依赖 winutil.exe 服务，用于模拟 Linux 下的目录环境，所以需要在 Windows 环境变量 `PATH` 中加入这个可执行文件的目录，该可执行文件可在 https://github.com/steveloughran/winutils 下载。

用户可自行选择 Hadoop 对应版本下载，假设下载的版本为 hadoop-2.6.3，可以通过图 A-1 所示配置环境变量。

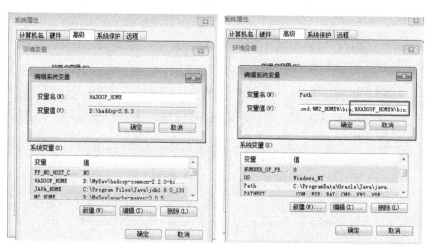

图 A-1　winutil.exe 环境变量配置

A.5 误删表数据

开发人员或者运维人员可能需要经常在测试环境与线上环境之间切换,如果将应该在测试环境执行的删除操作错误地执行到了线上环境,则可能对整个公司都是一个灾难。

如果整个表或者大量数据被删除了,此时依靠手动恢复显然不可行,那么只能通过最近的 SNAPSHOT 或者 Hadoop 备份进行恢复。如果少量数据被删除,则可以通过如下步骤手动恢复。

如果表开启了集群间复制,那么表属性 KEEP_DELETED_CELLS 应该已经被设置为 true,否则需要立即执行如下命令将该属性设置为 true,以防止 HBase 主压缩将删除的数据清理。

```
alter 's_behavior', { NAME => 'pc', KEEP_DELETED_CELLS => TRUE }, { NAME => 'pf', KEEP_DELETED_CELLS => TRUE }
```

接下来以 s_behavior 表 pc 列族数据被误删除为例,使用 HBase shell 命令来找回误删除数据,然后重新插入表,代码清单 A-7 描述了删除数据后通过时间区间查询出被删除的数据。

代码清单 A-7 HBase 找回数据

```
hbase(main):003:0> desc 's_behavior'
Table s_behavior is ENABLED
s_behavior
COLUMN FAMILIES DESCRIPTION
{NAME => 'pc', DATA_BLOCK_ENCODING => 'NONE', BLOOMFILTER => 'ROW', REPLICATION_SCOPE => '0', VERSIONS => '1', COMPRESSION => 'NONE', MIN_VERSIONS => '0', TTL => 'FOREVER', KEEP_DELETED_CELLS => 'FALSE', BLOCKSIZE => '65536', IN_MEMORY => 'false', BLOCKCACHE => 'true'}
{NAME => 'ph', DATA_BLOCK_ENCODING => 'NONE', BLOOMFILTER => 'ROW', REPLICATION_SCOPE => '0', VERSIONS => '1', COMPRESSION => 'NONE', MIN_VERSIONS => '0', TTL => 'FOREVER', KEEP_DELETED_CELLS => 'FALSE', BLOCKSIZE => '65536', IN_MEMORY => 'false', BLOCKCACHE => 'true'}
2 row(s) in 0.1020 seconds

hbase(main):004:0> alter 's_behavior', { NAME => 'pc', KEEP_DELETED_CELLS => TRUE }, { NAME => 'pf', KEEP_DELETED_CELLS => TRUE }
Updating all regions with the new schema...
1/1 regions updated.
Done.
Updating all regions with the new schema...
1/1 regions updated.
```

```
Done.
0 row(s) in 3.8360 seconds

hbase(main):005:0> scan 's_behavior'
ROW                                        COLUMN+CELL
 543210000000000000000922337050807043971 column=pc:o, timestamp=1528783696479,
 value=1002
 41
 543210000000000000000922337051331220088 column=pc:o, timestamp=1524550412164,
 value=1002
 61
 543210000000000000000922337051331220088 column=pc:v, timestamp=1523541994550,
 value=1002
 61
 543210000000000000000922337051331220088 column=pc:o, timestamp=1523541994550,
 value=1004
 62
 543210000000000000000922337051331220088 column=pc:v, timestamp=1523541994550,
 value=1004
 62
 543210000000000000000922337051331220088 column=pc:o, timestamp=1523541994550,
 value=1009
 63
 543210000000000000000922337051331220088 column=pc:v, timestamp=1523541994550,
 value=1009
 63
 543210000000000000000922337051331220092 column=pc:o, timestamp=1523541994550,
 value=1001
 10
 543210000000000000000922337051331220092 column=pc:v, timestamp=1523541994550,
 value=1001
 10
5 row(s) in 0.0880 seconds

hbase(main):007:0> deleteall 's_behavior', '543210000000000000000922337050807043
97141'
0 row(s) in 0.0390 seconds

hbase(main):008:0> scan 's_behavior'
ROW                                        COLUMN+CELL
 543210000000000000000922337051331220088 column=pc:o, timestamp=1524550412164,
 value=1002
 61
 543210000000000000000922337051331220088 column=pc:v, timestamp=1523541994550,
```

```
 value=1002
  61
 54321000000000000000092233705133l220088 column=pc:o, timestamp=1523541994550,
 value=1004
  62
 54321000000000000000092233705133l220088 column=pc:v, timestamp=1523541994550,
 value=1004
  62
 54321000000000000000092233705133l220088 column=pc:o, timestamp=1523541994550,
 value=1009
  63
 54321000000000000000092233705133l220088 column=pc:v, timestamp=1523541994550,
 value=1009
  63
 54321000000000000000092233705133l220092 column=pc:o, timestamp=1523541994550,
 value=1001
  10
 54321000000000000000092233705133l220092 column=pc:v, timestamp=1523541994550,
 value=1001
  10
4 row(s) in 0.0310 seconds

hbase(main):012:0> scan 's_behavior',{TIMERANGE =>[0,1528783696480] , raw => true}
ROW                      COLUMN+CELL
 *54321000000000000000092233705080704397l* *column=pc:o, timestamp=1528783696479,*
 *value=1002*
  *41*
 54321000000000000000092233705133l220088 column=pc:o, timestamp=1524550412164,
 value=1002
  61
 54321000000000000000092233705133l220088 column=pc:v, timestamp=1523541994550,
 value=1002
  61
 54321000000000000000092233705133l220088 column=pc:o, timestamp=1523541994550,
 value=1004
  62
 54321000000000000000092233705133l220088 column=pc:v, timestamp=1523541994550,
 value=1004
  62
 54321000000000000000092233705133l220088 column=pc:o, timestamp=1523541994550,
 value=1009
  63
 54321000000000000000092233705133l220088 column=pc:v, timestamp=1523541994550,
```

```
 value=1009
 63
 543210000000000000000922337051331220092 column=pc:o, timestamp=1523541994550,
 value=1001
 10
 543210000000000000000922337051331220092 column=pc:v, timestamp=1523541994550,
 value=1001
 10
5 row(s) in 0.0320 seconds
```